Advanced Steam Plant

Conference Organizing Committee

F Harris
(Retired) GEC ALSTHOM Turbine Generators Limited

J Jesson (Chair)
Mitsui Babcock Energy Limited

A Kirby
Consultant

J Muscroft
GEC ALSTHOM

A Strang
GEC ALSTHOM Turbine Generators Limited

A Tindall
National Power plc

IMechE
Conference Transactions

I MECH E
150th Anniversary
1847 - 1997

International Conference on

Advanced Steam Plant
– New Materials and Plant Designs and their Practical Implications for Future CCGT and Conventional Power Stations

21–22 May 1997

Organized by the Steam Plant Committee
of the Power Industries Division
of the Institution of Mechanical Engineers (IMechE)

Sponsored by

Co-sponsored by
The Institute of Energy
The Institute of Electrical Engineers
The Institute of Materials

IMechE Conference Transaction 1997 – 2

Published by Mechanical Engineering Publications Limited for
the Institution of Mechanical Engineers, Bury St Edmunds and London.

First Published 1997

This publication is copyright under the Berne Convention and the International Copyright Convention. All rights reserved. Apart from any fair dealing for the purpose of private study, research, criticism or review, as permitted under the Copyright, Designs and Patents Act, 1988, no part may be reproduced, stored in a retrieval system, or transmitted in any form or by any means, electronic, electrical, chemical, mechanical, photocopying, recording or otherwise, without the prior permission of the copyright owners. Reprographic reproduction is permitted only in accordance with the terms of licences issued by the Copyright Licensing Agency, 90 Tottenham Court Road, London W1P 9HE. *Unlicensed multiple copying of the contents of this publication is illegal.* Inquiries should be addressed to: The Publishing Editor, Mechanical Engineering Publications Limited, Northgate Avenue, Bury St. Edmunds, Suffolk, IP32 6BW, UK. Fax: 01284 704006.

© The Institution of Mechanical Engineers 1997

ISSN 1356-1448
ISBN 1 86058 097 1

A CIP catalogue record for this book is available from the British Library.

Printed by The Ipswich Book Company, Suffolk, UK.

The Publishers are not responsible for any statement made in this publication. Data, discussion, and conclusions developed by authors are for information only and are not intended for use without independent substantiating investigation on the part of potential users. Opinions expressed are those of the Author and are not necessarily those of the Institution of Mechanical Engineers or its Publishers.

Contents

Keynote Session

C522/030/97	Development of high-efficiency USC power plants in Denmark H J R Blum and J Hald	3
C522/029/97	Overview of advanced steam plant development in Japan K Miyashita	17

Steam Turbines

C522/001/97	Steam turbines for advanced steam cycles A N Paterson	33
C522/020/97	Advanced high-efficiency turbines utilizing improved materials B Scarlin	49
C522/032/97	Advanced steam turbines for modern power plants H Oeynhausen, A Drosdziok, and M Deckers	65

Materials for Steam Turbines

C522/007/97	New materials for advanced steam turbines R W Vanstone and D V Thornton	87
C522/008/97	Development of 9–12% Cr steels for all-ferrite steam turbine at target temperature of 650°C K Hidaka, Y Fukui, R Kaneko, and T Fujita	99
C522/012/97	Development of 11CrMoWCo heat resistant steel for fossil thermal plants M Ohgami, Y Hasegawa, H Naoi, and T Fujita	115
C522/004/97	Experience in the manufacture of steam turbine components in advanced 9–12% chromium steels M Taylor and D V Thornton	125

Boilers

C522/023/97	Benson boiler with vertical tube water walls – principles and advantages J Franke, R Kral, and E Wittchow	143
C522/005/97	The design of a 1000MW coal-fired boiler with the advanced steam conditions of 593°C/593°C K Sakai and S Morita	155

C522/025/97	Two pass boiler design for advanced steam conditions I R Torkington, M Upton, and F George	169

Materials for Boilers

C522/027/97	New steels for advanced coal fired plant up to 620°C E Metcalfe, W T Bakker, R Blum, R P Bygate, T B Gibbons, J Hald, F Masuyama, H Naoi, S Price, and Y Sawaragi	189
C522/026/97	Crack stability assessment for advanced 9CR steels in boiler components N Taylor, E Lucon, V Bicego, and P Bontempi	201

Emerging Technologies

C522/031/97	Emerging coal-fired power generation technologies D H Scott	213

Plant Operation

C522/024/97	Contributions of by-pass systems to the flexibility of advanced steam plants R R Rohner, U R Blumer, and J Aregger	225

Authors' Index   259

The Power to Serve Your Future.

In a rapidly developing world, meeting the ever growing demand for power generation will be one of man's biggest challenges.

Power generation is an essential part of our lives. It harnesses our natural resources, drives our manufacturing bases and helps to develop the technologies of the future.

As a leader in power generation, GEC ALSTHOM is building some of the most important power stations ever constructed. Power stations which will generate energy which is readily available, reliable and clean.

With our proven strength in technology, project management, financing and lifetime support, GEC ALSTHOM is helping to meet the challenge - helping to provide the power to serve your future.

GEC ALSTHOM Power Generation Division
38 avenue Kléber 75795 PARIS CEDEX 16 FRANCE
Tel: (33) 1 47 55 26 56 Fax: (33) 1 47 55 27 28

Related Titles of Interest

Title	Author	ISBN
Heat Transfer	IMechE Conference 1995–2	0 85298 950 4
Energy for the 21st Century	IMechE Seminar 1996–3	1 86058 035 1
Commissioning and Operation of Combined Cycle Plant	IMechE Seminar 1994–9	0 85298 933 4
Combined Power and Process – an Exergy Approach	F J Barclay	0 85298 942 3

For the full range of titles published by MEP contact:

Sales Department
Mechanical Engineering Publications Limited
Northgate Avenue
Bury St Edmunds
Suffolk
IP32 6BW
UK

Tel: 01284 763277
Fax: 01284 704006

Keynote Session

C522/030/97

Development of high-efficiency USC power plants in Denmark

H J R BLUM
Fælleskemikerne I/S Fynsværket, Denmark
J HALD
Elsam/Elkraft, c/o Department of Metallurgy, Technical University of Denmark, Denmark

The growing demand for clean electricity generation is bringing coal in a difficult position as the worlds largest and most reliable basic energy source. To maintain coal as a major basic energy source for electricity production efficiency improvement of coal fired power plants dependent on improved high temperature materials has been considered. Materials for a coal fired power plant with an efficiency of about 50% are now available, but during the design of such a plant special attention must be paid to superheaters and furnace panels. A European R&D project aiming at a "700°C" coal fired power plant partly based on the use of Ni-based alloys may raise efficiency to about 55% within the next decades.

BACKGROUND.

The total electricity generating capacity in the ELSAM system, covering the western part of Denmark including Jutland and Funen, is app. 5.900 MW. This consists of 4.200 MW large coal fired steam plants located at the major cities, 500 MW wind turbines and 1.200 MW small decentralized units fired with natural gas, biomass and/or municipal waste. Most of the units are equipped for combined production of electricity and district heating, and this increases the overall annual thermal efficiency in the ELSAM system to more than 60%. An overview of installed production capacity and of the annual generation of electricity and district heating in the ELSAM system related to basic energy sources, is presented in **fig. 1**.

After the oil crises in 1973 coal was chosen as the basic fuel for electricity generation. In 1980 all electricity produced in the Elsam system was based on coal and as coal was regarded to be a stable and reliable energy source, Elsam found it benificial from an economical point of wiev to improve the efficiency of future coal fired power plants. A long term strategy including the necessary materials R&D was set up to support the development of highly efficient coal fired power plants.

Later strong governmental demands for environmentally acceptable electricity production has instigated the Danish power companies to reduce the SO_x, NO_x and CO_2 emissions. Demanded reductions of SO_x and NO_x have been achieved by installation of flue gas cleaning systems. Concerning CO_2 our government has instructed Elsam firstly to install large capacity of wind turbines, secondly to stimulate the installation of a large number of small decentralized units based on gas, waste or biomass for combined heat and power production, and thirdly to fire about 1 mio. ton pr. year of the CO_2 neutral biomass comming as a surplus from agriculture. Together with the improvement of the thermal efficiency of fossil fired units this forms our present strategy for CO_2 reduction.

Wind turbines and small decentralized units using CO_2 neutral fuel for combined heat and power production are already widely exploited in the ELSAM system. But low production flexibility of decentralized heat/power units and unpredictability of production from wind turbines restricts the total useful output from such units. Therefore the larger proportion of electricity production in the future is still foreseen to be based on large centralized units based on fossil fuels including as much biomass as possible.

Natural gas and coal are currently the dominating fossil fuels used worldwide for electricity production. Advanced natural gas fired combined cycle plants, providing a thermal efficiency of more than 55%, is the cleanest way to produce electricity from fossil fuel with respect to SO_x, NO_x and CO_2 emissions. But the large amount of gas fired CC-plants already planned or commisioned, and the limited natural gas resources in Europe may influence the price and reliability of the future gas market. In Denmark coal is far the cheapest energy source, and it is also expected to be the most reliable energy source in the future with respect to long term resources and market stability. Since 1980 Elsam has been amongst the cheapest producers in Europe of electricity based on fossil fuel. Consequently ELSAM has focused on coal as the basic fossil fuel for the future.

COAL/BIOMASS BASED HIGH EFFICIENCY GENERATING CONCEPTS.

Intensive R&D work has been made and research programs are currently running within ELSAM to investigate the state of the art and the development potential of different coal based electricity generating concepts with respect to efficiency and emission control. Later these studies has also included biomass co-firing.

Three coal based technologies for power generation have been considered: IGCC-, PFBC-, and USC-plants. Comprehensive study projects have been made for these technologies with the clear goal of building power plants with a size of 300-400 MW if the results of the study project showed a reliable technical solution giving high efficiency at acceptable costs [1]. Basic results from these study projects are shown in **fig. 2**.

Later the capability of biomass co-firing was added to the concept studies. Co-combustion of coal and straw in existing coal fired conventional plants and in a CFB plant as well as in coal gasification test plants were made [2].

The overall conclusions from the study projects and co-firing tests were that the most efficient and technical feasible solution is the USC concept. Neither the IGCC- nor the PFBC concept showed better efficiencies or potentials. Furthermore the investment costs are much lower for the USC concept.

PAST DEVELOPMENTS OF USC PLANTS.

Basic obstacles to an increase in thermal efficiency of a fossil fired steam power plant are the limitations on achieveable steam parameters set by the creep properties of construction materials for thick section boiler and turbine components, and by the corrosion properties of superheater materials.

Since 1980 all new coal fired units in the ELSAM system have been supercritical plants with single reheat systems. Four 350-400 MW units were commisioned in the period 1983 to 1992. The designs of these units were based on conventional materials. Ferritic steels, ranging from plain C-steel up to the well known 12%Cr steel X20CrMoV121, were used for tubing and thick section components. Within the range of mechanical properties set by the given materials it was possible through an optimization of the plant concept and steam parameters to increase the thermal efficincy from 42% in the first units to 45% in the latest unit. These efficiencies are based on the lower heat rate but including flue gas cleaning. Steam parameters were 250 bar and 540°C to 560°C. [3]. Main data and choice of materials for these four units are given in **fig.3**.

With the development and ASME approval of the ferritic steel Grade 91 by the ORNL in USA in 1985, an increase of steam temperature and pressure up to app. 300 bar and 580°C was considered realistic. After intensive design studies of such a USC plant by ELSAM in collaboration with boiler manufacturers and turbine producers, two identical 400 MW USC units with steam parameters 290 bar and 580°C was ordered in 1992 for commisioning in 1997 and 1998. Combined with double reheat an efficiency of 47% is foreseen, including energy losses for flue gas cleaning [4]. Both units will be fitted for combined production of electricity and district heating as well as for pure condensation mode. In **fig.4** the basic data for the units are given.

FUTURE DEVELOPMENTS OF USC PLANTS.

In 1994 Elsam stated that when making full use of the recent materials development it would be possible to construct a coal fired USC power plant with steam parameters 325bar/620-630-°C leading to an efficiency of about 50% [5].

Since then a design study for a 400 MW coal fired USC unit with double reheat has been made in collaboration with a Danish boiler maker. Based on steam parameters 325 bar - 610/630/630°C the design should aim for a thermal efficiency of more than 50%.

Key components in such a unit for which special attention for design and choice of material has to be paid are:

- Thick section boiler components and steam lines.

- Turbine rotors and casings.

- Superheaters.

- Furnace panels.

Thick section boiler components and steam lines.
For thick section boiler components and steam lines the newly developed 9-12%Cr steels NF616 and HCM12A was chosen. In the international research project EPRI RP1403-50 pipe production, pipe bending and welding procedures have been developed and this has led to ASME Code approval of the steels in 1994 as P92 (NF616) and P122 (HCM12A) in Code Cases 2179 and 2180 [6]. In the second round of this project superheater outlet headers of NF616 and HCM12A have been produced and installed in one of the Elsam 400 MW USC plants with steam parameters 290 bar/580°C. This demonstrates that the materials are now ready for use in large scale projects. For the 9%Cr-steel NF616 610°C might be regarded as the maximum service temperature for constrution of furnace external boiler components. At higher temperatures the relatively high oxidation rate will limit the use of this material. With respect to oxidation HCM12A can be used at temperatures up to about 650°C because of its higher Cr-content.

Even stronger ferritic/martensitic materials are under development which in future may allow live steam temperatures up to app. 650°C. One cadidate is the steel NF12, which is a 12%Cr-W-Co alloyed steel currently under development by Prof. Fujita and Nippon Steel. Tentative estimates of creep rupture strength is about 100 MPa at $650°C/10^5$ h [7]. If these expectations can be verified by long term creep rupture data, a steel with the necessary mechanical strength for thick section components in a 325bar/630°C USC plant is available.

Turbine rotors and casings.
Similar to the thick walled boiler component materials, new steels for turbine rotors and casings have been developed in Japan and Europe. In Europe this development has been made in the cooperative COST 501 program, which has resulted in the commercialisation of new alloys of the type 10%Cr-Mo-W-V-Nb-N-(B) for turbine rotors and casings. Materials testing including long term creep rupture tests up to more than 50.000 h together with fabrication of a number of turbines demonstrate the reliability of these mateials [8,9].

In Japan similar development programs are in progress. New rotor steels like the Japanese HR1200 - where the introduction of Co to a 12%Cr-W steel not only increase the high temperature strength but also the oxidation resistance - are needed to have a material, which can be used for the construction of a turbine with steam parameters higher than 325bar/620°C [10].

Superheaters.
Selection of suitable materials for superheaters is a very delicate problem. With increasing steam temperature mechanically stronger materials with improved gas side corrosion and steam side oxidation resistance in the final section of the superheaters must be considered. High temperature corrosion in coal fired boilers may be a problem at surface temperatures above 620°C strongly depending of the coal quality. As the coal qualities used by ELSAM are imported from all over the world, our boilers must be able to accept a wide range of coal

qualities. Austenitic steels with 18%Cr or more and with improved mechanical properties is the answer to these demands.

To handle superheater materials problems a detailed program for the calculation of superheater tube life has been developed by Elsam. The calculations include effects on creep life of fireside corrosion and of the rising metal temperature due to steam side oxidation [11]. With this program it has been possible to verify the behavior of the chosen candidate materials. Furthermore the calculations indicate that the most important factors for the life of a superheater is the fireside corrosion rate and the internal heat transfer coefficient. This means that more attention must be paid to high temperature corrosion when choosing tube materials for secondary reheaters than for superheaters.

Candidate materials for superheaters and reheaters in future USC plants are shown in **fig.5**. The fine grained TP347HFG and the Super 304 developed by Sumitomo Metal Industries and Mitsubishi Heavy Industries [12,13] may be suitable for a 325bar/610-630°C USC plant if reasonable coal qualities are used. When using more aggressive fuels higher alloyed steels like HR3C [14] and NF709 [15] must be considered.

Most of the mentioned steels are tested in a high temperature superheater rig installed in power plant Vestkraft unit 3 as part of a Brite-Euram project [16]. The test rig allows steam temperatures up to 620°C and the scheduled maximum test time is 21.000 hours with an option for up to 40.000 hours. First part of the loop was dismantled last year after app. 8.000 hours exposure. Gas side corrosion and steam oxidation data as well as mechanical properties of base metal and welds, after long time exposure are the aims of this project.

Furnace panels.
Prediction of the effect of service exposure on the degradation of furnace panels and the selection of suitable materials is even more complicated than for superheaters. Media temperature is lower but heat flux is much higher and the oxide formation on the inside of the tube and the complex geometry comprising finned tubes is worse. Moreover a weld friendly steel which need no PWHT is mandatory. In a "conventional" supercritical boiler with steam parameters 250bar/540°C the necessary heat pick up in the furnace to balance the heat transfer from flue gas to water/steam in the whole boiler process maximizes the temperature of the water/steam in the furnace panels to 420°C at the outlet. The maximum temperature on the tube surface after 10^5 h of service does not exeed 500°C, even when the temperature increase due to oxide formation on the water/steam side is considered. Therefore low alloy ferritic steels like 1Cr½Mo can be used for furnace panel tubes, normally without any risk for severe degradation.

When increasing the steam parameters to e.g 325bar/610°C the heat pick up in the furnace must consequently be increased to balance the boiler process avoiding too hot flue gas at the inlet to the superheaters. Maximum water/steam temperature at the furnace panel outlet must be increased to 470°C. resulting in a calculated midwall tube temperature of about 510°C at start of service. But midwall temperatures due to the accelerated growth of magnetite on the internal tube surface will increase to about 550°C during service. Therefore a stronger and if possible more oxidation resistant material than 1Cr½Mo is needed.

Three newly developed steels have been selected as candidate materials for furnace panels in future boilers with advanced steam parameters. The high alloyed 12%Cr tube steel HCM12 developed by Sumitomo Metal Industries and Mitsubishi Heavy Industries [17] has excellent creep strength, oxidation and corrosion resistance and due to a duplex microstructure of app. 30% δ-ferrite and 70% tempered martensite it is possible to weld this steel without preheat and PWHT. The lower alloyed 2½Cr tube steel HCM2S developed by Sumitomo Metal Industries and Mitsubishi Heavy Industries [18] and the Mannesmann developed 2½%Cr tube steel 7CrMoVTiB1010 [19] both have sufficient high temperature strength and a metallurgy which makes it possible to omit PWHT. The chemical composition and mechanical properties for all three steels are given in **fig.6**.

Testing of HCM12, HCM2S and 7CrMoVTiB1010 is in progress in Denmark to establish practical experience with the handling of these steels. In the furnace panels of an existing subcritical once through boiler, test sections of all three steels have been installed for about two years, and service under cycling conditions have been tested. No problems have been encountered during the production of the test panels or during operation. This demonstrates the usefulness of these steels in future USC boilers.

In connection with the in-plant test a computer furnace panel calculation program has been set up to simulate the service exposure and life consumption of a furnace panel in a USC boiler during operation [20]. These calculations first of all demonstrate that the temperature rise during operation is strongly depending on the rate of self oxidation at media temperatures higher than app. 450°C and of the deposition rate of oxides comming from the feed water. High quality feed water chemistry assuring minimum oxide in feed water is therefore a must if USC plants with advanced steam parameters are realized.

Based on our calculations it can be demonstrated that the use of either HCM2S or 7CrMoVTiB1010 is sufficient for our next generation of coal fired USC plant, 325 bar/610° having media temperatures in the furnace panel lower than 470°C. If a higher media temperature should be considered better oxidation resistance is needed and HCM12 should be used.

As a curiosity it can be mentioned that in relation to Elsams activities concerning use of biomass, a special boiler with steam parameters 185 bar/540°C operating on straw and wood chops involving a construction of a furnace made of HCM12 cooled by steam at 500 - 530°C is under construction. This boiler will be a demonstration of the capability of HCM12 for extreme high temperature performance in furnace panels.

SUMMARY AND FUTURE ASPECTS.

Demands for environmentally acceptable electricity production has instigated ELSAM to improve the efficiency of its coal fired power plants. The latest generation of Elsams coal fired USC plants will be in operation next year having an efficiency of 47%. In combination with wind turbines, decentralized combined heat/power plants and intensive use of biomass, the gorvernmental demands for CO_2 reduction will be achieved.

A design study for a new improved coal fired 400 MW USC plant with an efficiency of more than 50% enabled by steam parameters 325bar/610°C/630°C/630°C has been carried out. The results of the design study show that it is possible to construct such a USC plant, making use of existing materials, but special attention must be paid to the design and material selction for superheaters and furnace panels.

Further improvements of the USC concept by rising steam parameters may be possible with the use of Ni-based alloys. A steam cycle operating at 375bar and 700°C has been considered ending up with an efficiency of about 55%. Within the EU a collaborative project is under consideration to study the feasibility of such a steam cycle. Combined with co-firing of coal and 30% biomass, the CO_2 emission for such a plant will end up at the same magnitude as for an advanced natural gas fired CC plant.

REFERENCES.

[1] Kjær, S. Koetzier, H. van Liere, J. Rasmussen, I:
New coal based power plant concepts - a comparison of efficiency, economy, environmental and operational aspects.
VGB-Konferenz "Fossilbefeuerte Dampfkraftwerke mit fortgeschrittenen Auslegungsparametern", 16-18 Juni 1993, Kolding, DK.

[2] Nielsen, C.: Increased Use of Biomass - the Danish Solution. 3rd Munich Discussion Meeting "Energy Conversion from Biomass Fuels", Oct. 1996.

[3] Lind-Hansen, A. Lindhart, S.: Der Block Esbjerg 3 - Erfarungen aus Inbetriebnahme und erster Betriebszeit.
VGB-Konferenz "Fossilbefeuerte Dampfkraftwerke mit fortgeschrittenen Auslegungsparametern", 16-18 Juni 1993, Kolding, DK.

[4] Kjær, S.: Kohlenstaubbefeurte Kraftwerksblöcke mit fortgeschrittenem Wasser-/Dampfprozess.
VGB Kraftwerkstechnik 70 (1990) H.3, S. 201-208.

[5] Blum, R.: Materials development for power plants with advanced steam parameters - Utility point of Wiev. Materials for Advanced Power Engineering 1994, Liege October 3-6, 1994.

[6] Metcalfe, E ed. Proc. The EPRI/National Power Conference: New Steels for Advanced Plant up to 620°C", London, May 1995.

[7] Naoi, H. Ohgami, M. Mimura, H. Fujita, T.: Mechanical properties of 12Cr-W-Co ferritic steels with high creep rupture strength.
Materials for Advanced Power Engineering 1994, Liege October 3-6, 1994.

[8] Berger, C. Mayer, K.H. Scarlin, R.B.: Turbinenkonstruktionen mit neuen Stählen für hohe Dampfparameter. Teil 1: Ergebnisse der Werkstoffentwicklung.
VGB-Kraftwerkstechnik 74 (1994), H.4 S.338-345.

[9] Engelke, W. Franc, J.C. Scarlin, R.B. Busse, L.: Teil 2: Herstellerspezifische Konstruktionsmerkmale.
VGB-Kraftwerkstechnik 74 (1994), H.4 S.346-360.

[10] Hidaka, K. Shiga, M. Nakamura, S. Fukui, Y. Kaneko, R. Watanabe, Y. Fujita, T.: Development of 12Cr steel for 650°C USC steam turbine rotors.
Materials for Advanced Power Engineering 1994, Liege October 3-6, 1994.

[11] Blum, R. Henriksen, N. Larsen, O.H.
Lifetime Evaluation of Superheater Tubes Exposed to Steam Oxidation, High Temperature Corrosion and Creep. Power Plant Technology 1996. Int. Conf. Kolding, Denmark 4-6 September 1996

[12] Teranishi, H. Sawaragi, Y. Kubota, M. Hyase, Y.: Fine-Grained TP347H Steel Tubing with High Elevated-Temperature Strength and Corrosion Resistance for Boiler Applications.
2nd International Conference on Improved Coal-Fired Power Plants. EPRI Palo Alto, USA, 1988.

[13] Sawaragi, Y. Ogawa, K. Kato, S. Natori, A. Hirano, S.: Development of the Economical 18-8 Stainless Steel (Super 304H) having High Elevated Temperature Strength for Fossil Fired Boilers.
The Sumitomo Search, No. 48, January 1992 p.50-58.

[14] Sawaragi, Y. Teranishi, H. Iseda, A. Yoshikawa, K.: The Development of new Stainless Steel Tubes with High Elevated Temperature Strength for Fossil Power Boilers and Chemical Plants.
The Sumitomo Search, No. 40, December 1990 p.146-158.

[15] Quality and Mechanical Properties of NF709 for Power Plant Boilers.
Technical publication, Nippon Steel Corporation, May 1993.

[16] Blum, R. Chen, Q. Scheffknecht, G. Vanderschaeghe, A.
A Project to Determine High-Temperature Corrosion Characteristics of Different Steam Generator Materials under Operating Conditions and Steam Temperatures up to 620°C.
VGB Kongress "Kraftwerke 1995", 5. bis 7. September 1995 Essen.

[17] Iseda, A. Sawaragi, Y. Teranishi, H. Kubota, M. Hayase, Y.: Development of New 12%Cr Steel Tubing (HCM12) for Boiler Application.
The Sumitomo Search, No. 40, November 1989, p.41-56.

[18] Masuyama, F. Yokoyama, T. Sawaragi, A. Iseda, A.:Development of tungsten strengthened low alloy steel with improved weldability.
Materials for Advanced Power Engineering 1994, Liege October 3-6, 1994.

[19] Bendick, W. Ring, M.: Stand der Entwicklung neuer Rohrwerkstoffe für den Kraftwerksbau in Deutschland und Europa. VGB-Konfernz "Werkstoffe und Schweisstechnik im Kraftwerk 1996" 8. - 9. Oktober 1996, Cottbus.

[20] Henriksen, N. Larsen, O.H. Vilhelmsen, T.: Lifetime Evaluation of Evaporator Tubes Exposed to Steam Oxidation, Magnetite deposition, High Temperature Corrosion and Creep in Super Critical Boilers.

Electricity production in West - Denmark

ELSAM + municipal generation companies

Fuel	Capacity	Production
Coal	4200 MW	~75 %
Gas Biomass Waste	1200 MW	20 %
Wind	500 MW	5 %

Coal, gas, biomass and waste fired power plants have an overall efficiency of more than 60 % due to cogeneration of electricity and district heating.

fig. 1

Coal-based High-efficient Electricity Generating Concepts

	USC	PFBC	IGCC
1990-1992	385 MW 290 bar/ 580°C/580°C/ 580°C $\eta = 47\%$	375 MW GT: 75 MW/900°C ST: 300 MW/ 185 bar/ 540°C/540°C $\eta = 44\%$	300 MW GT: 1160°C ST: 130 bar/ 540°C/540°C $\eta = 45\%$
Trend 2000	400 MW 325 bar/ 610°C/630°C/ 630°C $\eta \sim 50\%$	Increased steam parameters 460 MW GT: 870°C ST: 290 bar 580°C/580°C/580°C $\eta \sim 48\%$	General improvements 417 MW GT: 1260°C ST: 125 bar/ 510°C/510°C $\eta \sim 47.3\%$
> 2010	400 MW 375 bar/ 700°C/725°C/ 725°C $\eta \sim 55\%$	Increased GT inlet temperature by partial coal gasification $\eta \sim 50\%$	Further Improvement: Hot gas cleaning combined with SOFC $\eta \sim 53\%$

Fig. 2

Super critical coal fired plants in Elsam

Size	Year of Comm.	Steam data	Eff. cond.
2 x 350 MW	1983/1984	250 bar/540°C/540°C	42%
1 x 390 MW	1991	250 bar/540°C/540°C	44%
1 x 380 MW	1992	250 bar/560°C/560°C	45%

All units are sea water cooled and can operated either in 100% condensation mode or in co-generation of electricity and district heating.

Overall efficiency by co-generation ~85-90%

Construction materials:

    Boiler: C-steel, 1Cr1/Mo,2 1/4Cr1Mo, x20CrMoV 12.1 (560°C unit 347 H and 321 H)

    Steam lines: x20CrMoV 12.1

    Turbine HP, IP: 1CrMoV/560°C unit x20CrMoV 12.1

Fig. 3

USC units under construction

Fuel:	natural gas and coal
Size:	2 x 400 MW
Steam data:	290 bar/580°C/580°C/580°C
Efficiency:	47%

Construction materials:	
Boiler:	C-steels, 1Cr1/2 Mo, 2 1/4Cr1Mo, P91, 347HFG
Steam lines:	P91
Turbine:	10% Cr, Mo, V, Nb, N

Fig. 4

Materials for Superheater tubes

Chemical composition:

	Cr	Ni	Mo	Nb	Ti	Others:
TP347HFG	18	10		1		
Super 304H	18	9		0,4		Cu, N
NF709	20	25	1,5	0,25	0,05	N
HR3C	25	20		0,4		N

Creep rupture strength:

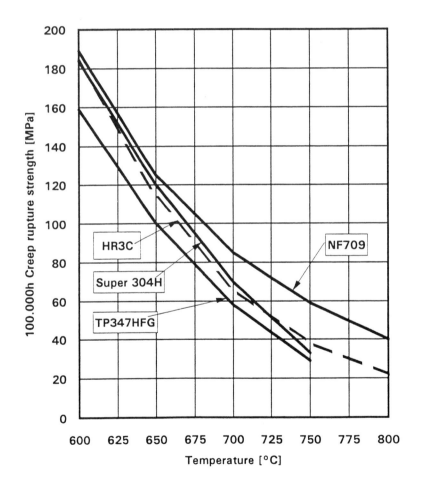

Figure 5

Materials for furnace panels

Chemical composition:

	C	Cr	Mo	W	Others:
1Cr½Mo	0,13	0,90	0,50		
HCM2S	0,06	2,25	0,30	1,60	V, Nb, N, B
HCM12	0,10	12,00	1,00	1,00	V, Nb
7CrMoTiB1010	0,07	2,40	1,00	-	V, Ti, N, B

Creep rupture strength:

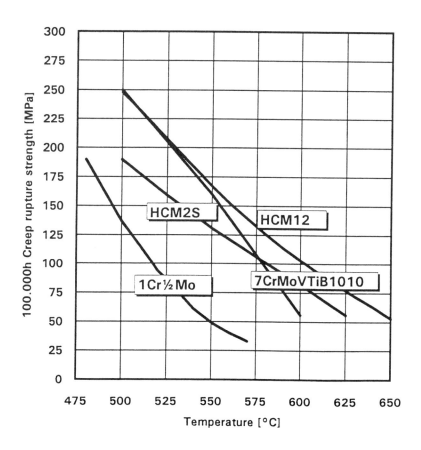

Figure 6

Overview of advanced steam plant development in Japan

K MIYASHITA BSc
Electric Power Development Company Limited, Tokyo, Japan

This paper gives an overview of development of advanced steam power plants in Japan. Conventional steam power plants completed recently or now being built or designed are adopting supercritical steam condition exceeding the 'standard' steam temperature of 538/566°C. This is the result of combined Japanese effort to realize ultra supercritical (USC) steam cycle. Effort is being made to realize further increase in the steam pressure and temperature to 30MPa and 630/630°C. Development of advanced gas turbines designed to inlet gas temperature of around 1500°C and combined steam cycle power plants is briefly reviewed.

1. INTRODUCTION

In Japan, energy supply policy is to divert from use of oil to alternative to oil energy source, and dependence on oil of the primary energy supply is reduced over the last 20 plus years from 77% in Fiscal 1973 (abbreviated hereinafter e.g. F1973) to 56% in F1995. Regarding the energy source for electricity, dependence on oil is markedly reduced over the last 20 plus years from 71% in F1973 to 18% in F1995.

Development of alternative energy to oil is to develop power stations utilizing either nuclear, natural gas or coal as the energy source. The shares in the electrical energy generated (kWh) by each of the energy source is 34% by nuclear, 22% by natural gas and 13% by coal in F1995.

The technical and management effort in the coal and natural gas generation field is reduction of investment cost for generating facility and reduction of generating cost (common to both generation types), further increase in plant efficiency of coal fired generation through increase in steam temperature and pressure, realization of economical gasification of coal and further increase in inlet gas temperature of gas turbines to achieve higher efficiency from combined cycle generation.

The reason why Japanese are investing effort in the development of advanced steam cycle and advanced gas turbine combined cycle is that Japan lacks domestic produce of primary energy source, thus has to depend on import. Efficient utilization of energy is essential. Advanced plants are economical compared to the existing plants in the rising fuel cost environment.

Sharp increase in worldwide energy consumption is estimated in the coming decades as a consequence of economic development in the Asian countries, especially in the newly industrialized economic sectors, Southeast Asia, and China and India where economy is growing rapidly. The growth in the consumption will accompany gradual rise in the price of raw material, food and energy resources.

In addition to economic reason, there is environmental consideration. Advanced plants contribute to no regret policy to mitigate the effect of increased emission of green house effect gases through lesser emission of carbon dioxide gas from coal fired power plants owing to higher efficiency and through lesser release and leakage of methane gas from gas production fields and liquefying plants due to lesser use of natural gas in the combined cycle power plants.

In Japan, gradual but comprehensive deregulation of economic activity is taking place and the electric power business is not exempted from this movement. This will have some bearing on research and development activity in the electrical power generation field in the near future.

2. ULTR SUPERCRITICAL STEAM CYCLE DEVELOPMENT

Development of supercritical steam cycle in the range 31.0-34.5MPa, 621/565/538-649/566/566°C was tried in the United States in Philo 6 and Eddystone 1 around 1957-1959. In Europe, supercritical steam cycle in the range 15-30MPa, 600-650°C was applied to small but many installations amounting to 14, mostly in the chemical industry, in Huels, Hatingen 3 & 4, Leverkusen 5 & 7 in Germany, etc., in 1950's to early 1960's. But the development had been either abandoned due to technical difficulties or because of economics. Recently in Japan, increase in steam temperature and pressure has commenced. Better understanding of metallurgy has resulted in availability of improved material which offer essentially the same handling as the existing material. Application of USC steam condition at or above 593/593°C at 24.1MPa is made to several large capacity units and a power station with 24.1MPa, 593/593°C, Matsuura power station unit 2 owned by EPDC (the Electric Power Development Co., Ltd), is under commissioning, being operated at 100% load in the satisfactory condition and will be put into commercial operation toward the end of June, 1997.

The development of ultra supercritical steam cycle with steam temperature above the almost standard supercritical temperature range of 538-566°C is the result of combined Japanese effort among utilities, steel manufacturers and boiler & steam turbine manufacturers. Preceding to the actual application, there have been research and development covering the entire range necessary in realizing the cycle. The effort ranges from development of material for the critical high temperature components to fabrication of their prototype, confirmation of material behavior under the service condition and so on.

2.1 USC development by EPDC

Manufacturers and EPDC have been collaborating in the pursuit of the ultra supercritical steam cycle as depicted in Figure 1, and in the effort till 1993 from around 1980 (termed as Phase 1), two conditions, regenerative with double reheat were the targets. One of them aimed at 31.4MPa, 593/593/593°C termed as Step I condition and the other at 34.3MPa, 649/649/649°C termed as Step 2 condition. In the current effort since 1994, the target condition has been adjusted, namely, regenerative with one reheat at 30.0MPa, 630/630°C (termed as Phase 2). It is to be noted that the effort has been subsidized by the Agency of Natural Resources and Energy in the MITI since 1982.

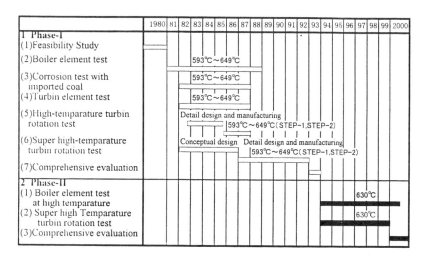

Figure 1 USC steam cycle development schedule by EPDC

(1) Phase 1

In the Phase 1 effort, **(a)** the candidate superheater tube material was installed in the boiler furnace as bypass panels from the last superheater header and steam is overheated up to 649°C. Testing is done for the duration of 42,838hrs for the 649°C heated portion and for 52,838hrs for the 593°C heated portion. The testing was performed in Takasago power station, which utilizes domestic coal as fuel, in collaboration with Mitsubishi (MHI).

In the parallel effort to the above, **(b)** superheater tube material test probes were inserted in the high temperature portion in the boiler furnace of Matushima power station to find out the corrosion resistance of the candidate material against outside surface oxidation in the environment of imported coal combustion and to compare with the one in the environment of domestic coal combustion at Takasago. Test was continued for 30,602hrs.

Also in Phase 1, **(c)** three tests to demonstrate the viability of USC steam turbines were performed.

    (i) One was heat soaking test of the HP inlet pressure retaining components performed in Takasago power station in collaboration with MHI.

    (ii) The second one was a long term high temperature rotation experiment of two partial models of rotor assemblies (one designed for 593°C and manufactured by an improved 12Cr forged material and the other one for 649°C and fabricated by the austenitic material A286) performed in Takasago P/S in collaboration with Hitachi and Toshiba. The former prototype underwent testing, mainly consisted of continuous operation for the duration of 4,818hrs to find out creep lifetime consumption, specifically in the area of highly stressed central hole. The latter underwent testing to find out the endurance of the rotor, specifically fatigue lifetime consumption of outside layer of the rotor due to temperature cycles. The test duration reached 2,161hrs and the start up frequency reached 151 times.

(iii) The last was the production of the prototype 1000MW turbines to both 593 & 649°C temperatures which were procured from MHI. The demonstration operation was performed to the duration of 14,302hrs for 593°C turbines and to 5,130hrs for 649°C turbine at Wakamatsu Coal Utilization Research Center of EPDC, the turbine being incorporated and used as the actual machine to generate electricity by the steam produced from 50MWe demonstration purpose atmospheric fluidized bed combustion boiler from 1986 through 1992. Referring to this FBC demonstration testing, the development effort has materialized as the 350MW large scale fluidized bed combustion boiler as a replacement to Takehara unit 2 oil fired boiler and is in satisfactory commercial operation since in service in June 1995.

The conclusion reached in Phase 1 is that 593°C class large capacity steam turbines consisted of improved ferritic material are judged feasible while 649°C class steam turbines of austenitic material construction are judged remote from early realization considering the stringent operational restraint imposed upon due to use of austenitic material with higher thermal expansion coefficient and lower thermal conductivity, higher lifetime consumption due to start up and shutdown caused by the same characteristics of the material as above and economic disadvantage.

(2) Phase 2

As a further effort to the realization of USC steam cycle, EPDC is presently undertaking Phase 2 USC steam cycle development, starting in 1994 and continuing until 2000. The following are the major areas of interest.

Development of Phase 2 USC material is performed including new candidate material. The list of material for the boiler is shown in Table 1. The list includes not only the material for high temperature use but also the ones for intermediate temperature use, the latter being included because of their possible contribution to reduction of boiler weight through their higher strength. High temperature property is confirmed along with steam corrosion characteristics, in furnace oxidation characteristics, etc. Specimens are inserted into unit 1 boiler of Takehara power station. Testing is done in collaboration with Hitachi, MHI and Ishikawajima-Harima (IHI). Full scale models of coils, water wall panels, headers and main piping are fabricated to confirm suitability of new material to fabrication process.

Development of high temperature material for the turbine is made with more emphasis on ferritic material. Two assemblies of a partial 1000MW class HP&IP turbine rotor which is connected in tandem and each embedded with first row blades is tested in the high temperature rotation facility shown in Figure 2. The candidate material for the main components are shown in Table 2. In order to obtain running hours exceeding the initial transition creep phase, an accelerated experiment at elevated temperature of 650°C for the duration of 500hrs at minimum is planned. The test is performed in collaboration with Hitachi and Toshiba.

Full scale steam valves (main steam valve and drain valve) and flanged connections for main steam leads and turbine casing are fabricated and leak tightness and integrity, etc. is being confirmed.

Table 1 Phase 2 USC material for boiler element test at Takehara P/S

Components		Material	Nominal Composition
Economizer Tubes		WT780C	0.8 Cr-0.5Mo-Cu
		Tempaloy HT780	0.6 Cr-0.3Mo-Cu
Waterwall Tubes		HCMV3	1.25Cr-1Mo-V
		NF1H	1.25Cr-1Mo-V-Nb
		Tempaloy F-2W	2Cr-0.5Mo-W-V-Ti
Superheater Tubes	Ferritic Material	NF616	9Cr-0.5Mo-1.8W-V-Nb
		HCM12A	12Cr-0.4Mo-2W-Cu-V-Nb
		Tempaloy F-12M	12Cr-0.7Mo-0.7W-Cu-V-Nb
		SAVE12	10Cr-3W-Co-Cu-V-Nb
		NF12	11Cr-2.6W-Mo-Co-V-Nb
	Austenitic Material	SUPER304H	18Cr-9Ni-3Cu-Nb-N
		Tempaloy AA-1	18Cr-10Ni-3Cu-Nb-Ti-N
		NF709	20Cr-25Ni-Mo-Nb-Ti-N
		SAVE25	23Cr-18Ni-3.5Cu-W-Nb-N
Header/Main Steam Pipes		NF616	9Cr-0.5Mo-1.8W-V-Nb
		HCM12A	11Cr-0.4Mo-2W-Cu-V-Nb
		Tempaloy F-12M	12Cr-0.7Mo-0.7W-Cu-V-Nb
		NF12	11Cr-2.6W-Mo-Co-V-Nb

Table 2 Phase 2 material for major components in high temperature steam turbine rotation test

Components	Material	Nominal Composition
Rotor	HR1200	11Cr-2.6W-0.2Mo-0.2V-0.1Nb-2.5Co-B
	SFR1200	10Cr-1.8W-0.7Mo-0.2V-Nb-3Co-B
Moving blades	12Cr forged steel	10.5Cr-2.6W-0.1Mo-0.2V-0.1Nb-0.2Re-1Co-B
	TAF650	11Cr-2.7W-0.2Mo-0.2V-0.1Nb-2.9Co-B
Inner casing	12Cr forged steel	10Cr-1.7W-0.7Mo-0.2V-3Co
	9Cr forged steel	9Cr-1M0-0.2V
	9Cr forged steel	9Cr-1.8W-0.5Mo-0.2V-0.1Nb-B
Bolts	12Cr forged steel	10.5Cr-2.6W-0.1Mo-0.2V-0.1Nb-0.2Re-1Co-B
	TAF650	11Cr-2.7W-0.2Mo-0.2V-0.1Nb-2.9Co-B

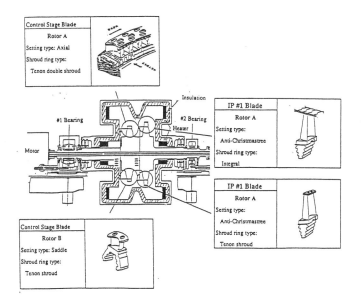

Figure 2 High temperature turbine rotation test assembly

2.2 High temperature material development

Development of high temperature material has been essential for the realization of ultra supercritical steam cycle. Significant progress has occurred in 9-12Cr steel and anstenitic steel in the last 10 plus years. Confirmation of suitability of such material to welding and other process necessary for the fabrication of final components is made. Prototype production and field testing in the actual service condition, e.g., in boilers or in high temperature rotation facility, etc. has enabled the selection of material for actual application.

Table 3 through 5 summarize the candidate high temperature material selected so far by the Japanese power plant manufactures as applicable to critical components in the USC steam cycle, namely, final superheater (SH) tubes and reheater (RH) tubes, SH / RH outlet headers and main steam / hot reheat steam pipes, turbine components such as high pressure (HP) and intermediate pressure (IP) steam turbine rotors, moving blades in the first row of HP & IP turbines and other important pressure retaing parts like main steam valves and HP & IP inner casings.

It is noted that creep rupture strength of new material has increased significantly compared with the material belonging to the same group according to composition of Cr-Mo or Cr-Ni as shown in Figure 3 (Reference 1). Understanding of strengthening mechanism and function of minor elements, etc. has altered the new material development altogether. In addition to traditional solid solution strengthening (addition of Mo) and precipitation strengthening (addition of V, Nb and Ti), control of Cr nitride and Cr carbide, addition of minor elements such as B and N, replacement of Mo by W, addition of Cu to austenitic steel and overall optimization of alloying elements, etc. are incorporated. The creep rupture strength has been increased by 50°C in general at the temperature which gives allowable stress of 49MPa.

Table 3 Material for final superheater/reheater tubes (heated surface)

Material	Composition	Remarks
NF709	20Cr-25NiMoTiNbBN	produced by Nippon Steel Corporation considerd applicable to 630°C USC
HR3C	25Cr-20NiNbN	produced by Sumitomo Metal Industries adopted as 600°C SH and 600/610°C RH tube considerd applicable to 630°C USC
SAVE25	23Cr-18Ni3.5CuWNbN	produced by Sumitomo Metal Industries new material, property to be clarified
Tempaloy A-3	22Cr-15NiNbBN	produced by NKK Corporation considerd applicable to 630°C USC
Super304H	18Cr- 9Ni3CuNbN	produced by Sumitomo Metal Industries adopted as 600°C SH and 600/610°C RH tube considered applicable to 630°C USC
Tempaloy AA-1	18Cr-10Ni3CuNbTiN	produced by NKK Corporation new material, property to be clarified
ASME TP347HFG	18Cr-10NiNb	TP347H fabricated to ASME code case 2159 fine grained and strengthened
Tempaloy A-1	18Cr-10NiTiNb	produced by NKK Corporation adopted as 600/610°C RH tube
NF616	9Cr-0.5Mo1.8WVNb	produced by Nippon Steel Corporation applicability to be clarified
HCM12	12Cr-1Mo1WVNb	produced by Sumitomo Metal Industries applicability to be clarified

Table 4 Material for final SH/RH headers & main steam/hot reheat steam pipes

Material	Composition	Remarks
SUS316HTP	16Cr-12Ni2.5Mo	applicable to steam temperature approaching 650°C USC
HCM12A	12Cr0.4Mo2WCuVNb	produced by Sumitomo Metal Industries adopted as 600/610°C final SH/RH headers and main/hot reheat steam pipes applicable to 630-650°C USC
NF12	11Cr2.6WMoCoVNb	produced by Nippon Steel Corporation applicability to be clarified
Tempaloy F-12M	12Cr0.7Mo0.7WCuVNb	produced by NKK Corporation applicability to be clarified
SAVE12	10Cr3WCoCuVNb	produced by Sumitomo Metal Industries applicability to be clarified
NF616	9Cr0.5Mo1.8WVNb	produced by Nippon Steel Corporation applicable to 600°C USC
HCM12	12Cr1Mo1WVNb	produced by Sumitomo Metal Industries* applicable to 600°C USC
ASME SA335 P91	9Cr1MoVNb	developed by ORNL for fast reactors adopted as 600/610°C final SH/RH headers and main/hot reheat steam pipes
ASME SA387 Gr91	9Cr1MoVNb	developed by ORNL for fast reactors adopted as 600/610°C hot reheat steam pipe and RH headers

Table 5 Material for USC steam turbine components

Components	Material	Composition	Remarks
High pressure/ Intermediate pressure rotor	Modified 12Cr forging	11Cr1W1MoVNbN	adopted by Toshiba for 600/610°C rotors
	SFR1200	10Cr1.8W0.7MoVNb3CoB	going to be checked at 630°C
	HR1100	11Cr1.2Mo0.4WNbV	adopted by Hitachi for 600/600°C rotors
	HR1200	11Cr2.5W0.2Mo0.2V2.5Co	going to be checked at 630°C
	TMK-1	10Cr1.5Mo0.2VNbN	adopted by MHI for 600/600°C rotors
	TMK-2	10Cr0.3Mo2W0.2VNbN	adopted by MHI for 600/610°C rotors
Moving blade (first row, HP/IP turbine)	New 12Cr	10.5Cr2.6W0.1MoVNbReCoB	adopted by Toshiba for 600/610°C blades applicable to 630°C
	HR1100	11Cr1.2Mo0.4WNbV	adopted by Hitachi for 600/600°C blades
	TAF650	11Cr0.2Mo0.2V3.0W3.0Co	going to be checked at 630°C
	R-26	austenitic refractory alloy	adopted by MHI for 600/600°C blades
Nozzle box Inner casing	12Cr cast steel	10CrMoVNbN	adopted by Toshiba for 600/610°C inner shell
	12Cr or 9Cr forging		adopted by Toshiba for 600°C nozzle box
	CrMoVB cast steel		adopted by Hitachi for 600°C HP inner shell
	9Cr forging		adopted by Hitachi for 600°C nozzle box
	12Cr cast steel		adopted by Hitachi for 600°C IP inner shell
	MJC-12	12Cr cast steel	adopted by MHI for 600/600°C nozzle chamber and inner casing
Main steam stop valve/ Control valve / CRV	New 12Cr casting	10Cr1.8W0.7MoVNb3CoB	adopted by Toshiba for 610°C CRV's applicable for 630°C casing and valve
	New 12Cr forging	10Cr1.8W0.7MoVNb3CoB	going to be checked at 630°C
	9Cr forging	9Cr1MoNbV	ASTM A182F91 adopted by Toshiba for 600°C MSV and CV's adopted by Hitachi for 600/600°C MSV,CV and CRV's adopted by MHI for 600/610°C MSV,GV,ICV and RSV's

The strengthening of some of the new material exceed 50°C as shown in Figure 3. It is noted that either Mo content is replaced by addition of W in the case of HCM12A and NF616 or Cu is added as a novel alloy element to the austenitic material.

Creep rupture strength alone is not sufficient for the selection of material for the actual usage. Other property such as fracture toughness, adaptability to fabrication and manufacturing process, endurance in the service environment, etc. has to meet the requirement.

This paper will not discuss findings made by researchers and boiler and steam turbine designers about such research into dependence of material wastage at elevated temperature under existence of SO2 upon Cr content, dependence of inner surface steam corrosion upon grain size, etc., which is an important information for the design of critical components.

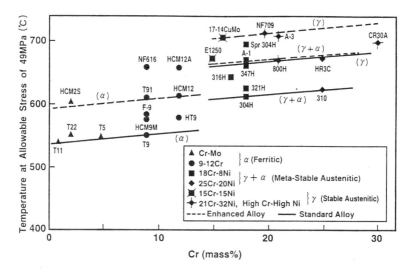

Figure 3   Temperature corresponding to allowable stress of 49Mpa of existing and new material as a function of Cr content

3. ULTRA SUPERCRITICAL STEAM CYCLE POWER PLANTS IN JAPAN

3.1 USC steam cycle power plants

Conventional steam power plants which have come into commercial service since 1990 or which will be in service prior to or around the year 2000 in Japan are summarized and listed in Table 6. It is noted that Kawagoe unit 1 & 2, which are twin 700MW natural gas fired units owned by Chubu Electric Power Company and came into service in 1989 and 1990 respectively with steam conditions of 31.0MPa, 566/566/566°C followed by Tsuruga unit 1, a 500MW coal fired unit owned by Hokuriku Electric Power Company and put into service in 1991 with steam conditions of 24.1MPa, 566/566°C have been the forerunners of an increase in the steam pressure and temperature from virtually the 'standard' steam conditions of 24.1MPa, 538/566°C for the preceding 30 years. When the units in Table 6 are classified into several groups according to the steam pressure and temperature, it results in Table 7.

Application of USC steam cycle improves plant efficiency as illustrated in Figure 4 compared with the plants with the prevalent steam conditions of 24,1MPa, 538/566°C.

In the design of the power stations being built according to the latest arts of the day, the following are the typical features.

a. Sliding pressure control is adopted because of high partial load efficiency and possibility of large capacity coal fired units being subjected to middle load operation since fairly large share of the power grid in Japan is occupied by nuclear power stations.

b. Environmental protection is an important role for planning power stations in Japan, and recent example of the NO x level in the exhaust gas from the coal fired units as agreed upon with the local governments is in the range 135-145ppm at 6% O_2 at the boiler outlet as a maximum under steady state operation and it is further required to be equipped with flue gas denitrification plants of 90% efficiency rather than that of 80%.

c. Flue gas desulfurisation plants are requirements in planning the coal fired units. So far, plants based on wet limestone gypsum system have been the selection. However, in the case of build & scrap replacement of twin domestic coal fired 265MW units of Isogo P/S owned by EPDC into twin 600MW units on the limited area of 12ha, a charcoal desulfurisation system is employed for the first time in the commercial power generation in Japan because of the limited space requirement by the system.

Table 6   Conventional steam cycle power plants in service in 1990 through around 2000 in Japan

name of station	main steam condition		R/H cond'n	output	owner utility	in service	B'r supplier	T/G supplier
	pressure	temperature						
Shiriuchi 2	24.1MPa	566°C	566°C	350MW	Hokkaido EPC	1997	IHI	Toshiba
Noshiro 1	24.5	538	566	600	Tohoku EPC	1993	B-Hitachi	Fuji
Noshiro 2	24.1	566	593	600	Tohoku EPC	1994	IHI	Toshiba
Haramachi 1	24.5	566	593	1000	Tohoku EPC	1997	MHI	Toshiba
Haramachi 2	24.5	600	600	1000	Tohoku EPC	1998	B-Hitachi	Hitachi
Higashi oogishima 2	24.1	538	566	1000	Tokyo EPC	1991	IHI	MHI/ME
Hirono 4	24.1	538	566	1000	Tokyo EPC	1993	IHI	Toshiba
Hitachinaka 1	24.5	600	600	1000	Tokyo EPC	2002	B-Hitachi	Hitachi
Kawagoe 2	31	566	566/566	700	Chubu EPC	1990	MHI	Toshiba
Hekinan 1	24.1	538	566	700	Chubu EPC	1991	MHI	Toshiba
Hekinan 2	24.1	538	566	700	Chubu EPC	1992	B-Hitachi	Hitachi
Hekinan 3	24.1	538	593	700	Chubu EPC	1993	IHI	MHI/ME
Tsuruga 1	24.1	566	566	500	Hokuriku EPC	1991	MHI	Toshiba
Nanaoota 1	24.1	566	593	500	Hokuriku EPC	1995	B-Hitachi	MHI/ME
Nanaoota 2	24.1	593	593	700	Hokuriku EPC	1998	IHI	Toshiba
Tsuruga 2	24.1	593	593	700	Hokuriku EPC	2000	MHI	Toshiba
Nanko 1	24.1	538	566	600	Kansai EPC	1990	MHI	MHI/ME
Nanko 2	24.1	538	566	600	Kansai EPC	1991	B-Hitachi	Hitachi
Nanko 3	24.1	538	566	600	Kansai EPC	1991	IHI	Toshiba
Maizuru 1	24.1	593	593	900	Kansai EPC	2003		
Maizuru 2	24.1	593	593	900	Kansai EPC	2003		
Misumi 1	24.5	600	600	1000	Chugoku EPC	1998	MHI	MHI
Tachibanawan	24.1	566	593	700	Shiokoku EPC	2000	B-Hitachi	Toshiba
Reihoku 1	24.1	566	566	700	Kyushu EPC	1995	IHI	Toshiba
Reihoku 2	24.1	593	593	700	Kyushu EPC	2001	MHI	Toshiba
Gushigawa 1	16.6	566	538	156	Okinawa EPC	1994	KHI	Hitachi
Gushigawa 2	16.6	566	538	156	Okinawa EPC	1995	B-Hitachi	MHI
Mastuura 1	24.1	538	566	1000	EPDC	1990	B-Hitachi	MHI
Mastuura 2	24.1	593	593	1000	EPDC	1997	B-Hitachi	MHI
Tachibanawan 1	25	600	610	1050	EPDC	2000	IHI	Toshiba/GE
Tachibanawan 2	25	600	610	1050	EPDC	2001	B-Hitachi	MHI
Isogo 1	25	600	610	600	EPDC	2002	IHI	Fuji
Shinchi 1	24.1	538	566	1000	Sohma J't PC	1994	B-Hitachi	Hitachi
Shinchi 2	24.1	538	566	1000	Sohma J't PC	1995	MHI	Toshiba

IHI:Ishikawajima Harima Heavy Industries   B-Hitachi:Babcock Hitachi   MHI:Mitsubishi Heavy Industries
ME:Mitsubishi Electric   KHI:Kawasaki Heavy Industries   GE:General Electric   Fuji:Fuji Electric

Table 7   Ultra supercritical steam cycle power plants in Japan

main steam pressure	main steam temperature	hot reheat steam temperature	number of units	unit belonging to the group
	538°C	566°C	XX	not referred to
	538	593	1	(1)
24.1-25.0	566	566	3	(2)
MPa	566	593	4	(3)
	593	593	6	(4)
	600	600	3	(5)
	600	610	3	(6)
31.0MPa	566	566/566	2	Kawagoe 1/2

name of tha units
(1) Hekinan 3
(2) Tsuruga 1   Reihoku 1   Shiriuchi 2
(3) Nosiro 2   Nanao oota 1   Haramachi 1   Tachibanawan
(4) Matsuura 2   Nanao-oota 2   Tsuruga 2   Reihoku 2   Maizuru 1/2
(5) Misumi 1   Haramachi 2   Hitachinaka 1
(6) Tachibanawan 1/2 Isogo 1

d. Replacement project in Isogo offers another opportunity for employment of new technology in Japan. A tower boiler of IHI supply which is hanged from top girder structure which rests on the four corner columns of reinforced concrete construction is employed in unit 1 boiler because of limited space again. Tower boilers are popular in Europe. It is different from the ones in Europe that parallel path arrangement of coils is adopted in the downstream side of the convective heat transfer area.

e. Fuji electric is going to supply USC steam turbine for unit 1 of Isogo in cooperation with Siemens AG. High temperature material developed in COST (European Cooperation in the Field of Scientific and Technical Research) 501 project would be utilized in the project.

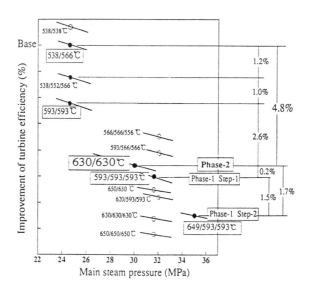

Figure 4   Plant efficiency improvement by applying USC steam cycle

4. ADVANCED GAS TURBINE AND COMBINED STEAM CYCLE POWER PLANTS

After long incubation period since pioneer GTCC power plants were installed around the turn of 1960's to 1970's in the U.S., the advancement of gas turbine technology toward higher inlet gas temperature and larger capacity is remarkable. Overall efficiency of 1300°C class gas turbines and combined conventional steam cycle which have been adopted in fair numbers in the power network far exceeds that of advanced conventional steam cycles. It would be not misleading to state that GTCC generation will continue to exist as a primary power generating system for the foreseeable future, unless economic advantage is severely impaired by rise in the price of natural gas due to demand / production gap, etc. or amount of natural gas permissible for consumption is restricted to control, for instance, release of methane gas from gas fields to certain amount worldwide in order to mitigate the effect of green house effect gases

The GTCC power plants in Japan, which were put into service in 1990 or after or which will be in service prior to or around year 2000 are listed in Table 8.

Table 8   Gas turbine combined cycle power plants in service in 1990 through around 2000 in Japan

name of station	inlet gas temperature	gas turbine type	gas turbine capacity		steam temperature H.P	steam temperature IP / LP	steam turbine capacity		station formation
Higashi niigata 4	1450°C	701G1	270MWx4		566°C	566/ - °C	265MWx2		2GT-1ST
Yokohama 7	1288	F9FA	225	4	538	538/275	125	4	1GT-1ST
Yokohama 8	1288	F9FA	225	4	538	538/267	125	4	1GT-1ST
Chiba 1	1350	701F	242	4	538	538/250	118	4	1GT-1ST
Chiba 2	1288	F9FA	233	4	538	538/262	127	4	1GT-1ST
Kawagoe 3	1288	F7FA	151	7	538	538/250	85	7	1GT-1ST
Kawagoe 4	1350	501F	158	7	538	538/ -	85	7	1GT-1ST
Shin nagoya 7	1288	F7FA	158	6	538	538/ -	85	6	1GT-1ST
Himeji No.1 5	1350	501F	139	3	538	538/ -	253	1	3GT-1ST
Himeji No.1 6	1204	F7FA	140	3	538	538/250	250	1	3GT-1ST
Yanai 1	1104	F7EA	78	6	485	- /163	42	6	1GT-1ST
Yanai 2-1, 2-2	1260	F7F	125.4	2	538	- /158	72.6	2	1GT-1ST
Yanai 2-3, 2-4	1260	F7F	125.4	2	538	538/ -	72.6	2	1GT-1ST
Shin ooita 1	1085	F7E	76	6	480	- /158	39	6	1GT-1ST
Shin ooita 2	1350	501F	144.4	4	538	- /161	73.1	4	1GT-1ST
Shin ooita 3-1	1288	F7FA	160	3	538	538/ 25	85	3	1GT-1ST

name of station	station capacity	supplier GT/ST	owner utility	in service
Higashi niigata 4	1610MW	MHI/MHI	Tohoku EPC	1999,2000
Yokohama 7	1400	GE/GE	Tokyo EPC	1998
Yokohama 8	1400	GE/GE	Tokyo EPC	1998
Chiba 1	1440	MHI/MHI	Tokyo EPC	2000
Chiba 2	1440	GE/GE	Tokyo EPC	2000
Kawagoe 3	1652	Hitachi/Hitachi	Chubu EPC	1997
Kawagoe 4	1650	MHI/MHI	Chubu EPC	1997
Shin nagoya 7	1458	Toshiba/Toshiba	Chubu EPC	1999
Himeji No.1 5	670	MHI/MHI	Kansai EPC	1995
Himeji No.1 6	670	Hitachi/Hitachi	Kansai EPC	1996
Yanai 1	700	Hitachi/Hitachi	Chugoku EPC	1990
Yanai 2-1	350	Hitachi/Hitachi	Chugoku EPC	1994
Yanai 2-2	350	Hitachi/Hitachi	Chugoku EPC	1996
Shin ooita 1	690	Hitachi/Hitachi	Kyushu EPC	1991
Shin ooita 2	870	MHI/MHI	Kyushu EPC	1995
Shin ooita 3-1	735	Hitachi/Hitachi	Kyushu EPC	1998,2004

It is evident that gas turbine performance is improving in a very rapid pace, that 1300°C gas turbines and combined cycle systems are extensively adopted, and that employment of multiple trains of GTCC is made for the large capacity GTCC power stations due to Japanese situation where natural gas is introduced from overseas in liquified form thus centralized utilization is inevitable.

As referring to advanced GTCC development, there are two groups' efforts, namely of MHI and of GE and its manufacturing associates.

4.1 Efforts by MHI for the advanced GTCC power plants

In Japan, MHI is developing 1500°C class air cooled and steam cooled (vanes and blades) combustion gas turbines (501G for air cooled 60Hz, 701G1 & G2 for air cooled 50Hz and H-series for steam cooled machines) and these are listed in Table 9. The first machine of 501G series is scheduled to start operation in the first quarter of 1997, and the first machine of 701G series is scheduled for loaded shop tests in the fourth quarter of 1997 (Reference 2).

Table 9 Advanced gas turbines under development by MHI

turbine blade cooling	air cooled		
frame	501G	701G1	701G2
speed(rpm)	3,600	3,000	3,000
G/T output(MW)	230	255	308
G/T efficiency(%)	38.5	38.7	39
combined output(MW)	343.3	373.9	454
combined efficiency(%)	58	57	58
pressure ratio	19	18.5	21

The turbine inlet gas temperature is 1500°C at the combuster outlet. 701G1 is developed for use in the Higashi niigata power plant unit 4. The compressor is a 17 stage axial flow machine whose pressure ratio is 19. The gas turbine has four stage.

For the turbine blades of rows 1 & 2, directionally solidified precision cast CM247LC is used. For the turbine blades of rows 3 & 4, MGA1400 developed by MHI / Mitsubishi Steel Manufacture is used. Rows 3 & 4 blades are shrouded while rows 1 & 2 blades are not shrouded. Vane rows 1 through 4 are precision cast using MGA2400.

The row 1 blade is cooled by the flow through the serpentine flow passage with angled turbulaters. Shower head cooling and full coverage film cooling are also employed. Thermal barrier coating is applied on the blade and platform surfaces. Impingement cooling is applied on the inside for the row 1 vane which differs from the blade cooling system.

The combuster is low NOx type but dry without water injection. In order to achieve NOx level of 21ppm at 16% O_2, almost all the combustion air is premixed with fuel gas and introduced into the primary zone of the combuster so that the flame temperature is kept at nearly the same level as that of the 1350°C class turbines, and the flame temperature is in the range of 1500-1600°C. Steam cooling system developed for the first time in the world is employed for the combuster cooling in order to reduce the cooling air.

The combined cycle plant efficiency is expected to be over 58%.

4.2 GE and its MA's effort in the advanced GTCC power plants

In Japan, General Electric and its manufacturing associate (MA) companies, Hitachi and Toshiba, is preparing for the advanced gas turbine combined power plant project. GE is participating along with Westinghouse in the stationary heavy duty turbine development in the advanced gas turbine system development program in the U.S., sponsored by DOE.

Advanced GTCC project being prepared for in service in the year 2002 is based on adopting high temperature turbines with inlet gas temperature at the first blade inlet of 1450°C. Refer to Table 10,

Table 10 Advanced gas turbines under development by GE/MA's

turbine blade cooling flame	air cooled MS7001G	steam cool MS7001H	air cooled MS9001G	steam cool MS9001H
speed(rpm)	3,600	3,600	3,000	3,000
G/T output(MW)	240		282	
G/T efficiency(%)	39.5		39.5	
combines output(MW)	350	400	420	480
combined efficiency(%)	58	60	58	60
pressure ratio	23	23	23	23

The combusters are of low NOx premix lean combustion type and in order to achieve NOx level of 21ppm at 16% O_2, similar diluted combustion as adopted in the gas turbines developed by MHI is employed. 12 combusters is adopted for 60Hz machine and 14 planned for 50Hz machine.

18 stage axial compressor is utilized along with four stage gas turbine. For the steam cooled series, steam cooling of rows 1 & 2 vanes and blades are employed. Material for row 1 vanes and blades are single crystal of Ni base superalloy and material for row 2 through 4 blades are directionally solidified precision cast material of GTD111 (Ni base superalloy).

Blade and vane are applied with thermal barrier coating.

The combined efficiency is expected at around 60% for steam cooled type and 58% for air cooled type.

5. DISCUSSION

It is a worldwide tendency that deregulation and any kind of reduction of public utility fare is favored irrespective of long term stability or reliability consideration. In order to clarify rationality of further increase in the steam temperature and pressure from the level now being achieved, it is necessary for Japanese utilities to examine economics of USC under higher plant efficiency, estimated higher fueling cost escalation and additional investment under decreasing component and installation cost environment. Discrepancy between actual operating and expected performance has to be carefully monitored in evaluating the effectiveness of new technology by plant operators.

Rapid increase in the gas turbine inlet temperature of stationary heavy duty gas turbines for generation is being made parallel to the speed in the aerospace combustion gas turbines. However economics of maintaining the 1300°C class advanced machines and reliability of the machines is not well known and awaits evaluation.

ACKNOWLEDGMENT

The author would like to express sincere gratitude to all those involved in the research & development of USC cycle and those being involved in the realization of the USC plants. Thanks are extended to those who assisted in the preparation of this paper.

REFERENCES

1. T. Yokoyama, F.Masuyama 'Application of boiler materials for ultra high temperature and high pressure power plants' 'The Thermal and Nuclear Power' Volume 45, No.11, pp43-54
2. S. Aoki et al. 'Development of the next generation 1500°C class advanced gas turbine for 50Hz utilities' International Gas Turbine and Aeroengine Congress & Exhibition, Birmingham, U,K. June 10-13, 1996

Steam Turbines

C522/001/97

Steam turbines for advanced steam cycles

A N PATERSON BSc, PhD
Power Generation Division, GEC ALSTHOM, Rugby, UK

SYNOPSIS

Substantial improvements in the efficiency of the reheat steam cycle can only come from increases in steam inlet pressure and temperature.

This paper demonstrates that much of the extensively proven steam turbine technology for conventional steam cycles is applicable to the design of turbines for advanced steam cycles. By incorporating both improved materials and improved internal steam conditioning to accommodate higher steam inlet temperatures and pressures, the proven technology can be substantially repeated with corresponding assurances of availability.

1. INTRODUCTION

Large power plants with advanced steam cycles were first ordered more than 40 years ago (Table 1).

Table 1. Some advanced steam plant ordered up to 1960

STATION	COUNTRY	UNIT OUTPUT MW	STEAM PRESSURE bar	STEAM TEMP. °C/°C/°C
HÜLS	GERMANY	85	294	600/560/560
EDDYSTONE 1	USA	325	346	649/565/565
DRAKELOW 12	UK	375	241	593/565

The early operating experience was disappointing, particularly with the boilers, due mainly to the application of austenitic materials to those pressure containing parts in contact with the highest temperature steam. As a result advanced steam cycles, here defined at the turbine as main pressures of 240 bar and above with main and reheat temperatures of 565°C and above, have since been avoided by utilities who have preferred relatively larger unit sizes with more modest steam cycles.

The original incentive for adopting advanced steam cycles was the large economic benefit of the efficiency improvements. In contemporary terms, for example, each 1% relative efficiency improvement on a coal fired 680 MW machine produces an estimated lifetime fuel saving of about $ US 10 million. Today this incentive is strongly reinforced by urgent environmental considerations. For example, the same efficiency improvement also produces an estimated life time reduction in CO_2 emission of about 0.8 million tonnes.

Substantial efficiency improvements with the steam cycle can only come from a return to advanced steam cycles. Relative efficiency improvements of 6% are attainable with the single reheat cycle by raising steam pressure to 300 bar and temperature to 600°C rather than the conventional values of 180 bar and 540°C most commonly specified today.

This paper is concerned only with those parts of the steam turbine influenced by advanced steam cycles. Turbines for service at 3000 rpm are used to illustrate the technology which, however, is also directly applicable to 3600 rpm turbines. After reviewing some turbine implications of advanced steam cycles, the main body of the paper shows that, by applying as necessary recently developed high alloy ferritic materials, the existing well tried technology provides a sound base on which modern GEC ALSTHOM steam turbines for advanced steam cycles are designed.

2. ADVANCED STEAM CYCLES

2.1 Efficiency Improvements

The thermodynamic efficiency of the conventional single reheat cycle can be improved by any means which increases the average temperature at which heat is added to the cycle.

For example, as main steam pressure is increased it becomes possible to heat feedwater to higher temperatures by tapping steam from within the turbine, significantly improving cycle efficiency. Optimisation of final feedwater temperatures should include the boiler as it affects boiler efficiency.

However, the most direct way of increasing the average temperature of heat addition to the cycle is by increasing both main and reheat steam temperatures. This improves relative efficiency by a rate of about 1.0% per 20°C rise over a wide range of temperature and pressure, for both single and double reheat (Fig.1). This improvement is virtually independent of the turbine design.

For low subcritical pressures, increasing main steam pressure also produces a large efficiency improvement (Fig.1). As the main steam pressure is increased, the average temperature of heat addition to the cycle increases but at a rate reducing with increased pressure.

In addition to these thermodynamic considerations, as steam inlet pressures are increased the steam volumetric flows correspondingly reduce. Consequently within both the HP and IP bladepaths, the expansion losses and the leakage and endwall losses tend to increase. This occurs because blade heights shorten corresponding to the reduced volumetric steam flow and blade widths increase to withstand the higher steam loading. These losses are determined by the seal design, bladepath aerodynamic design and manufacturing accuracy. Efficiency improvements arising from increased steam pressure are therefore to some extent dependent on the design and manufacture of the turbine.

Increasing main steam pressure also increases the level of steam wetness in the LP cylinder, causing a small loss in LP expansion efficiency.

The combined effect of all of these factors is that there is a level of steam supply pressure beyond which a further pressure increase produces no further efficiency improvement. This level will be lower for smaller machines which are more sensitive to blade end leakage losses than large machines.

Double reheating further improves relative efficiency by about 2% by increasing the average temperature of heat addition and by reducing LP steam wetness, even though there is some increase in expansion and leakage losses due to the reduction in steam flow relative to single reheat cycles. Also for double reheat, the level of steam supply pressure for the peak efficiency is higher than for single reheat.

2.2 Output and Pressure Capability

Standard turbine frames are designed for a maximum output and a maximum steam pressure at a given temperature, resulting in a maximum volumetric flow capacity.

For the same steam volumetric flow, frame sizes designed for higher maximum pressures are slightly larger, normally having one or two more stages for the increased expansion duty and wider stage pitching to accommodate strengthened blade path components. Similarly the higher maximum pressures require strengthened pressure containment components such as casings and bolting. Strengthening of blade path and pressure containment components uses existing technology and improved materials. Two of these standard frames will be used in the next section to illustrate the GEC ALSTHOM design principles for turbines for advanced steam cycles.

2.3 Temperature Capability

The limitation to increases in steam inlet temperature is the high temperature strength of the turbine and boiler materials.

For both forged and cast major turbine components, low alloy 1CrMoV steel has been extensively used for temperatures up to 565°C, beyond which there is a rapid reduction in the creep strength after long term service.

However, high alloy 9-12Cr ferritic steels have now been fully developed and are entering service for both forged and cast major turbine components with creep strengths suitable for long term service at temperatures up to 600°C. Variants of these 9-12Cr steels after short term testing show potential for long term service at temperatures of 625°C and possibly 650°C. The 9-12Cr steels also have a high temperature proof strength superior to the 1CrMoV steels (Reference 1,2).

Under transient operating conditions, high strain fatigue strength limits the allowable surface cyclic strain range. Compared with 1CrMoV, the 9-12Cr steels have much higher thermal fatigue strength (consistent with the better combination of creep and proof strength), a thermal expansion coefficient about 15% less, and a thermal diffusivity which is about 25% less. The allowable rapid temperature change of the rotor surface on a hot start is mainly controlled by the thermal expansion coefficient and thermal fatigue strength and is therefore much higher for the 9-12Cr material. For cold starts, the allowable rate of surface temperature rise is also strongly dependent on the diffusivity, resulting in comparable cold start rates at higher temperatures for the 9-12Cr material.

To retain bolted horizontal casing joints under conditions of high pressure and high temperature requires bolting material with much better high temperature creep relaxation properties than the high chrome material normally used for lower steam conditions. The nickel based Nimonic 80A material is excellent for this purpose and has substantial successful service experience.

3. ADVANCED SINGLE REHEAT

3.1 Configuration

Standard low reaction turbine designs for conventional steam conditions are inherently suitable for development to more advanced steam conditions and therefore no substantial changes to the configuration of the standard turbine cylinders and steam chests are required for advanced frames.

To outline most clearly the GEC ALSTHOM development logic for advanced single reheat turbines, this section reviews the designs of two standard frames with maximum ratings of 900 MW at 240 bar/565°C/565°C and 1100 MW at 300 bar/600°C/600°C respectively and relates them to the design of a standard frame with maximum rating of 680 MW at 180 bar/540°C/540°C. These three frames have approximately the same volumetric flow capacity but progressively increasing maximum steam pressure and therefore maximum output, together with progressively increasing maximum steam temperature.

The development of the more advanced frames essentially involves a number of detailed modifications to the highest temperature components, taking advantage of 9-12Cr materials where appropriate, together with greater emphasis on internal steam conditioning.

3.2 Steam Chests and Loop Pipes

For steam conditions of 240 bar/565°C/565°C and above, 9-12Cr materials are used for all steam chests and loop pipes (Tables 2, 3).

Table 2 Materials of HP major components

STEAM (bar/°C/°C)	180/540/540	240/565/565	300/600/600
LOOP PIPES STEAM CHEST	2.25%CrMo 1%CrMoV	9%CrMoVNbN 9%CrMoVNbN	9%CrMoVNbN 9%CrMoVNbN
CYLINDER			
OUTER CASING INNER CASING INNER BOLTS NOZZLE BOX ROTOR	2.25%CrMo 1%CrMoV 11%CrMoVNbN 1%CrMoV 1%CrMoV	2.25%CrMo 1%CrMo 11%CrMoVNbN 11%CrMoVNbN 1%CrMoV	2.25%CrMo 9%CrMoVNbN Nimonic 80A 11%CrMoVNbN 10%CrMoVNbN

Table 3 Materials of IP major components

STEAM (bar/°C/°C)	180/540/540	240/565/565	300/600/600
LOOP PIPES STEAM CHEST	2.25%CrMo 1%CrMoV	9%CrMoVNbN 9%CrMoVNbN	9%CrMoVNbN 9%CrMoVNbN
CYLINDER			
OUTER CASING INNER CASING INNER BOLTS HEAT SHIELD ROTOR	2.25%CrMo 1%CrMoV 11%CrMoVNbN - 1%CrMoV	2.25%CrMo 1%CrMoV 11%CrMoVNbN - 1%CrMoV	1%CrMoV 9%CrMoVNbN 11%CrMoVNbN 9%CrMoVNbN 10%CrMoVNbN

3.3 HP Cylinders

The construction of the HP cylinder for the 900 MW frame (Fig.2) is identical to that of the 680 MW frame but the cylinder is slightly longer and there is an extra stage for the increased enthalpy drop.

Steam is tapped from before the last stage, as for the standard design, for conditioning the casing interspace and a small axial flow is maintained by drawing the flow into the main gland (Fig.3). The flow is circumferentially uniform to preserve conditions of thermal axisymmetry on the outside of the inner casing and on the inside of the outer casing.

Main steam enters the cylinder through thermally sleeved connections, each sealed to a separate nozzle box located inside the inner casing. The use of separate nozzle boxes enables sliding pressure, throttle governing or nozzle governing to be employed without change to the basic cylinder construction.

The use of robust low reaction blading permits a substantial pressure drop across the first HP nozzle blades, with a correspondingly large reduction in the steam temperature as a result of the expansion. The pressure reduction directly reduces the pressure load on the inner casing and therefore reduces its thickness and the size of bolting required. Maximum use is made of the corresponding steam temperature reduction by designing the first row of moving blades to have a sufficient degree of negative reaction to cause recirculation of the cooler steam from the space after the first disc through the space between the nozzle boxes and the inner casing to the space before the first disc and into the main gland (Fig.3). This flow of cooler steam conditions local heat transfer coefficients and cools the inside of the inner casing and the associated bolting permitting further reductions in casing thickness and in the bolt size necessary for resisting both the reduced direct pressure and the reduced thermal moment loads. It also greatly assists in maintaining the thermal axisymmetry of the inner and outer casings.

The first stage disc of the rotor is also cooled by this flow of conditioning steam. The blading for this stage can either be the standard design for throttle governed or sliding pressure applications, or a specially designed control stage matched to the intended partial admission control duty. In either case pinned root fixings and integral tip shrouding are retained for maximum reliability. The combined effect of all these detailed design precautions is that conventional low alloy steels are retained for all the HP cylinder major components, except those in direct contact with the inlet steam (Table 2).

The HP cylinder for the 1100 MW frame is correspondingly larger than for the 680 MW frame with two extra stages for the increased duty. The same form of triple casing cylinder construction and steam conditioning is employed but, to accommodate the more advanced steam conditions, 9-12Cr steel is used for the inner casing together with Nimonic bolts (Table 2). The rotor is a monobloc 9-12Cr forging for the best possible integrity. The excellent high temperature properties of the 9-12Cr steels more than compensate for the increased temperature with respect to both steady state and transient operating conditions.

The journals of high chrome ferritic rotors are surfaced with a low alloy weld deposit to produce the best bearing characteristics. Two bearings per rotor are retained for all advanced HP and IP turbine designs which powerfully inhibit shaft vibration arising from either mass unbalance or instability. The two bearings between each rotor strongly decouple the vibration characteristics of adjacent rotors. Sufficient axial space is left between adjacent bearings to ensure that the load sharing between them is not significantly affected by the small bearing movements that are expected in service. Tilting pad journal bearings are used for their low power loss and good damping properties.

The inner and outer casings are slightly thicker to accommodate the higher pressure but the high temperature strength of the Nimonic bolts permits slim horizontal joint flanges to be retained, with all the advantages of rapid access for maintenance without loss of operational flexibility.

The diaphragm materials are 9CrMo and 12CrMo permitting the stage pitching to be minimised due to the improved high temperature material strength. Also small radial clearances between the diaphragms and rotor can be maintained since the high chrome diaphragm materials have a similar thermal expansion coefficient to the 9-12Cr rotor material.

To maximise cycle efficiency by increasing final feedwater temperature, an HP interstage steam tapping may be required. In this case a well proven ring sealed pipe interconnection between the inner and outer casings is used which does not disturb the standard pressure and temperature balance between the casings.

3.4 IP Cylinders

As for the HP cylinder, the IP cylinder arrangement of the 900MW (Fig.4) is similar to that for the standard 680 MW frame but the size is slightly larger with one extra stage for the increased duty. The single flow steam path is retained because it has the best expansion efficiency and, as the rotor hub diameter is smaller, it also has better thermal flexibility than double flow steam path designs.

Steam from the first tapping conditions the intercasing space, and cooling steam is injected into the space below the first nozzle blades to cool the first rotor disc and the main gland (Fig.5). The rotor material is improved to 9-12Cr (Table 3) and the diaphragm materials changed to high chrome ferritic materials.

For the more advanced 1100 MW frame, two extra stages are required for the single flow IP steam path relative to that for the standard 680 MW frame. Here steam from the second IP tapping can be used to condition the casing interspace, limiting the temperature and pressure to which the outer casing is exposed. The first tapping is extracted through a sealed pipe connection.

A 9-12Cr complete heat shield or nozzle box is used for containing the IP inlet steam. This shield, and the outer ring of the first diaphragm, both protect the inside of the inner casing from direct contact with flowing steam at reheat temperature, limiting the mean temperature level and temperature gradient through the casing. Similarly, the shield limits the outside temperatures of the main gland.

When necessary, to provide cooling for the second disc, an additional supply of cooling steam can be injected through the second stage diaphragm.

4. ADVANCED DOUBLE REHEAT

4.1 Configuration

The configuration of double reheat impulse steam turbines is mainly dictated by the design of the IP expansion. If the IP volumetric flow is sufficiently large to require a double flow IP cylinder, then the corresponding VHP and HP single flow expansions are in separate single flow cylinders. For lower volumetric flows, the single flow HP and IP expansions are placed in opposition in a combined HP/IP cylinder and the VHP remains as a separate single flow cylinder. An important turbine of this second type is used in the following review to outline the GEC ALSTHOM development logic for advanced double reheat turbines.

4.2 The SKAERBAEK/NORDJYLLAND 412 MW Turbines

4.2.1 Background

Two 412 MW double reheat machines have been ordered from GEC ALSTHOM for the SKAERBAEK and NORDJYLLAND power stations in Denmark. The SKAERBAEK machine is due to enter service in 1997 with NORDJYLLAND one year later. With advanced steam conditions (285 bar/580°C/580°C/580°C) and low condenser pressure (23 mbar due to seawater cooling temperature of 10°C), these machines will be the most efficient in the world operating on the steam cycle with an efficiency improvement of about 10% relative to the standard single reheat machine with a condenser pressure of 50 mbar. The turbines can also accept a large district heating load.

Layout and operating modes

This advanced turbine has 5 cylinders (Fig.6)

· one single flow VHP cylinder

· one combined HP/IP cylinder

· one assymmetrical double flow IP cylinder (associated with

  the district heating)

· two double flow LP cylinders (each with 1050mm long last stage blades).

There are two journal bearings per rotor and the separate thrust bearing is located in the second pedestal.

There are three pairs of steam chests, each pair controlling one of the three steam flows from the boiler. Each chest is spring-mounted on the foundation and connected to the turbine by flexible loop pipes.

In full condensing mode, with a nominal power output of 412 MW, the valves in the LP crossover pipes are open wide and the steam from each flow of the assymmetrical double flow IP cylinder exhausts vertically upwards from the ends of the cylinder and is fully expanded in separate double flow LP cylinders (Fig.6). In full district heading mode, the LP crossover valves are shut and the steam from each flow of the assymmetrical double flow IP cylinder, apart from a small quantity for LP cylinder ventilation, exhausts vertically downwards from the ends of the cylinder. This steam passes to two separate district heaters, providing a nominal heating load of 450 MJ/s while a nominal power output of 320 MW is produced by the steam as it expands through the first three cylinders,. For intermediate district heating loads the valves in the LP crossover pipes can modulate in an intermediate position to control the steam pressures to the district heating.

Only the VHP and HP/IP cylinders and associated steam valves are reviewed here. Since the pressures and temperatures are similar to that for the 1100 MW single reheat frame already described in the previous section, similar construction principles and material selections are used.

4.2.2 Steam chests and loop pipes

All the steam chests and loop pipes are of 9-12Cr steel. The VHP steam chests use a self sealing autoclave joint for the chest covers, avoiding any heavy bolting. The valve spindles are guided close to the valve heads which are themselves shaped to minimise buffeting while they are controlling the high energy steam flow. A similar design is used for the HP steam chests.

4.2.3 VHP Cylinder

The single flow expansion is in a separate cylinder (Fig.7) which permits a small rotor hub diameter of only 600mm with low centrifugal stress and low thermal inertia with a high critical speed and good shaft vibration characteristics. The rotor discs are narrow relative to their pitch and have large fillet radii between them and the rotor hub, minimising rotor surface stress concentrations. This low diameter design allows the longest blades for maximum expansion efficiency and the smallest diameters of interstage glands for minimum steam leakage. Due to the low axial thrust on the rotor produced by impulse blading, the balance piston diameter is only 640mm with correspondingly low thermal inertia and low steam leakage characteristics.

If conventional welded construction were used for the early diaphragms, the nozzles of the early diaphragms with large pressure drops would require to be very wide for diaphragm strength, with correspondingly reduced efficiency because of their poor aspect ratio. The early diaphragms are therefore of bridge type construction, where the structural strength is provided by a few wide radial members and narrow efficient nozzle blading is slotted between the rings, all of 9CrMo material. Later diaphragms are of standard spacerband welded construction using 12CrMo material.

The low rotor diameter also minimises the diameters of the triple casing construction, limiting the pressure loading on the casings and permitting them to be of minimum thickness. The inner casing is of 9-12Cr material 120mm thick and the outer casing of 2.25CrMo material 120mm thick.

Nimonic 80A bolting is used close to the inlet section of the inner casing, together with austenitic bolt collars to compensate for the low expansion coefficient of the 9-12Cr casing relative to the Nimonic bolts. The turbine is designed for full arc admission, with the two semi-circular nozzle boxes interconnected by sealed rings at the horizontal joints.

All the previously described provisions for steam conditioning the casing interspaces are fully implemented and the material improvements applied (Table 2).

4.2.4 HP/IP Cylinder

The HP and IP blade paths are both single flow, mounted in opposition to balance the small opposite axial thrusts in a single, compact combined HP/IP cylinder (Fig.8). The nominal steam supply pressures to each expansion are 74 bar and 19 bar respectively. The volumetric flows are therefore much higher than for the VHP cylinder and, with lower blade widths permissible because of lower steam loading, good blade aspect ratios allow high expansion efficiencies.

Also because of the lower steam pressures, a lower shaft critical speed is allowable and permits a hub diameter of only 640mm with a centre gland diameter of 680mm. The same advantages as for the VHP cylinder therefore apply with respect to low rotor thermal inertia and low interstage steam leakage.

A small amount of cooling steam is introduced through the main gland to cool the centre gland region and the first stage discs of both flows of the rotor which is 9-12Cr material. Both HP and IP inlets are protected by 9-12Cr complete heat shields, reducing the maximum temperatures of the inner casing of 9-12Cr material. Steam taken from before the last stage of the HP flow is used to condition the interspace between the casings before being returned to HP exhaust. The outer casing is of 1CrMoV steel, well suited to sustain the relatively high HP exhaust pressure and temperature.

All diaphragms are of welded spacerband construction using 9CrMo material for the early stages and 12CrMo materials for the remainder.

5. CONCLUSION

Relative to the subcritical steam cycles which currently dominate the international steam turbine market, the advanced steam cycles for which plant is available today permit large relative efficiency improvements of about 6% (single reheat) or 8% (double reheat).

In the case of steam turbine plant, machines for the advanced steam cycles substantially use the same turbine technology already extensively proven in service with conventional steam cycles, at both a turbine configuration level and a turbine component level. It is mainly the use of improved high chrome ferritic materials which have not only higher creep strength but also attractive thermal characteristics for design which permit this repetition of the technology in the high temperature cylinders. Conditioning steam flow practices, already widely used for standard high temperature cylinders, are also fully applied.

Conseqently the turbine availability, for the advanced cycle turbine plant is expected to be as high as for modern conventional steam turbine plant.

The good thermal properties of the high chrome ferritic materials also ensures equivalent operating flexibility, with allowable rates of turbine temperature change similar to current conventional machines.

For power generation using coal fuel in particular, the steam cycle remains the obvious choice. Boiler designs using improved materials are available to generate steam reliably at advanced steam conditions matching the steam turbine plant described in this paper. The trend towards advanced steam conditions, already evident in some countries, is therefore expected to extend as economic incentives and environmental constraints increase.

6. REFERENCES

1. THORNTON, D.V., VANSTONE, R.W. "New materials for advanced steam turbines", Advanced Steam Plant Conference, IMechE London, 1997

2. TAYLOR, M., THORNTON, D.V. "Experience in the manufacture of steam turbine components in advanced 9-12% chromium steels", Advanced Steam Plant Conference, IMechE London, 1997.

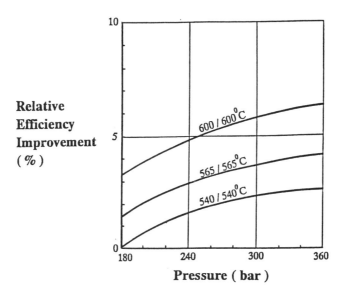

Fig 1 Effect of steam cycle on efficiency (single reheat)

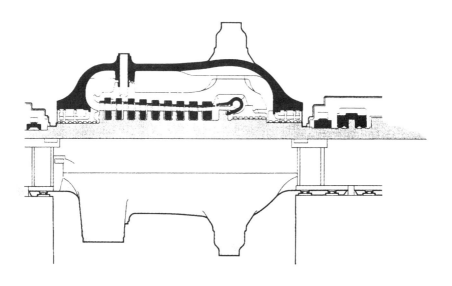

Fig 2 900 MW single reheat HP cylinder

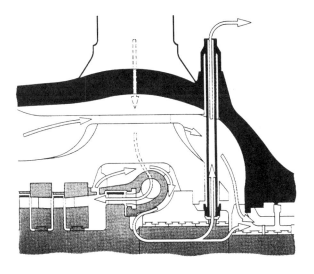

Fig 3   900 MW single reheat HP cylinder steam conditioning

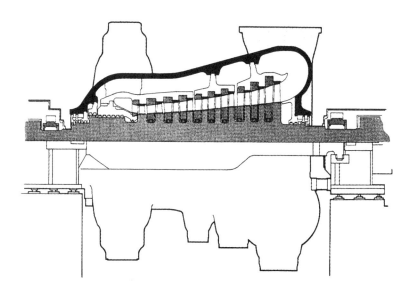

Fig 4   900 MW single reheat IP cylinder

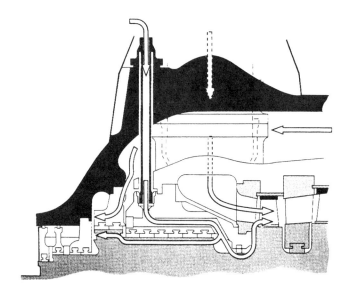

Fig 5   900 MW single reheat IP cylinder steam conditioning

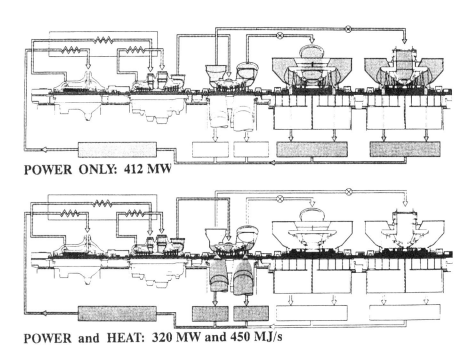

Fig 6   SKAERBAEK double reheat turbine showing operating modes

Fig 7　　SKAERBAEK VHP cylinder

Fig 8　　SKAERBAEK HP/IP cylinder

C522/020/97

Advanced high-efficiency turbines utilizing improved materials

B SCARLIN
Steam Turbine Development Department, ABB Power Generation Limited, Baden, Switzerland

SYNOPSIS

Economic considerations have been the basic driving force behind the development of high-efficiency advanced steam plant. The clearest benefits on the turbine side are obtained by increasing steam temperature. To permit this improvement materials of the 9 to 12 % Cr steel class with increased creep strength were required. In the last 10 years understanding of microstructural stability has been greatly improved and materials with increased temperature capability have been developed and introduced.

Appropriate designs are chosen to take best advantage of improved properties. Specifically designs using welded rotors and shrink-ring inner casings permit thermal transient stresses to be minimised. This concept of new materials and proven designs has now been applied.

Steam power plant with inlet temperatures up to 600°C can now be considered state-of-the-art and further efforts are being made in 2 directions:
- increase to about 650°C using steels which are currently under development
- increase beyond 700°C making considerable use of Nickel-based alloys.

1. INTRODUCTION

Recent analyses of predicted demand for electrical energy and known fuel reserves have shown that there will be a continued heavy demand on coal as primary energy source to satisfy growing requirements. In particular countries such as India and China, which have a rapidly growing demand for electrical power, have major reserves of coal. In addition coal and brown coal will still be burnt in European power stations, either to encourage our coal-mining industries or to make use of low-cost imported coal. Coal will remain the major source of fuel for power stations. Until reliable and low-cost methods of coal gasification are available, coal-burning steam power plant will remain world-wide the major source of electricity production.

Since there is a strong driving force towards increasing efficiency of electricity generation along with the desire to reduce environmental emissions, studies have been performed to determine what is the optimum overall design of a steam power plant. In particular it is seen that the optimum in terms of increased efficiency at minimum cost of the power plant is achieved by raising the temperature, and to a lesser extent also the pressure, of the live and reheat steam. The introduction of a second steam reheat may also be beneficial but is considerably more expensive for a corresponding efficiency improvement. In the same way an increase in the temperature of the feedwater entering the boiler, through more intensive preheating of the feedwater by extracted steam, increases the overall plant efficiency, but significantly increases costs. This is because more expensive materials and manufacturing procedures have to be introduced for the corresponding pipework.

Fig.1 illustrates, for a typical steam power plant, the extent to which the heat rate can be improved through raising the steam temperature and pressure. For example, a temperature rise from 560 to 600°C provides a relative efficiency increase of about 2 %. Increase in operating pressure also contributes about 0.5 % for an increase of 50 bar. On this basis, for new commissioned plant, operating efficiency values have been rising by an average of about 0.5 % per year for the last decade.

At the same time great importance is also placed on the reliability of the power plant, since lack of availability could rapidly cancel out the positive effect of increased efficiency and thereby reduced fuel costs. In this respect the statement by Kjaer (1) that supercritical power plant have been seen to be just as reliable as earlier subcritical plant is particularly important. For this reason the basic premise in the trend towards higher steam parameters and higher efficiency has been to retain known and proven turbine design principles and make use of improved materials. The materials have been developed, tested and validated over the last 10 years. They are now successfully in operation in a number of conventional steam and combined cycle power plants in Europe.

2. MATERIALS DEVELOPMENT

2.1 Issues Limiting Development and Use of Materials

The use of a material at elevated temperatures is generally limited by its creep strength and resistance to degradation, generally in the form of embrittlement or softening. This in turn is related to the microstructural stability. The general requirements for the present application are the following:

- high creep strength at a temperature of about 600°C
- high toughness and resistance to embrittlement during long-term use at high temperature
- resistance to steam oxidation and, in the case of boiler components, also fireside corrosion
- ease of fabrication for large forged and cast components, including weldability.

Development activities of this kind are frequently carried out through co-operation between a number of organisations, since it is possible in this way to share the costs and benefits. In addition it is possible to produce a much broader database of information, on the basis of which standardisation and general acceptance can more easily be achieved. In particular the

European COST Programme (CO-operation in Science and Technology) has been successful in developing and validating such steels, some of which are now in service. Participants in the COST programme have been the major European turbine and boiler manufacturers, producers of forgings, castings and pipes, electric power utilities and testing institutes.

For example, for rotor forgings the following targets were set in the COST programme:

- uniformity of mechanical properties in forgings of up to 1200 mm in diameter
- a creep strength at 600°C and 100 000 hours of 100 MPa
- toughness and low cycle fatigue properties better than current materials

The general approach began with an evaluation of the properties of presently available materials and a study of the development work currently being performed. Particular attention was paid to the understanding of the mechanisms of strengthening of steels in the 9 to 12 % Cr class, since a good combination of creep strength and oxidation resistance can be expected.

2.2 Alloy Development and Metallurgy of 9 to 12 % Cr Steels

Steels with a content of 9 to 12 % Cr provide a good combination of creep strength and oxidation resistance in steam. The best possible combination of mechanical properties can be achieved through quenching to produce a practically fully martensitic microstructure and then tempering the steel to precipitate carbides, thereby increasing toughness to the required level. Nevertheless the microstructures of these steels are not fully stable and further precipitation reactions occur, some of which are beneficial and some deleterious to the mechanical properties.

Alloying of steels is complex. Some elements are added specifically on account of the beneficial effects they have whereas others, such as sulphur, phosphorus, silicon and the other tramp elements (Sn, Sb, As, etc.), are kept to the minimum level which can be economically achieved on account of their negative effects, generally on toughness. Beneficial effects are provided by the following:

Chromium at a level between 9 and 12 % provides a good combination of creep strength and oxidation resistance. Higher Cr levels cause the appearance of low-toughness δ-ferrite.

Carbon at a level of 0.1 to 0.2 % is added to form carbides (mostly of chromium) which provide strength after heat treatment. Higher C levels may cause cracking during welding.

Molybdenum additions serve to form stable carbides and also contribute to solid-solution strengthening. Levels of 0.5 to 1.5 % are generally used.

Tungsten also makes a major contribution to solid-solution strengthening, but levels above 1 % can lead to the rapid formation of a coarse Laves phase, which can be detrimental to overall strengthening.

Niobium at a level of about 0.06 % forms a small volume fraction of carbides which are extremely stable at the austenitising temperature and hence prevent grain growth and consequent loss of toughness.

Vanadium is added at a level of about 0.20 to 0.25 % and forms nitrides with the nitrogen dissolved in the alloy. These vanadium nitride particles (also referred to as MX precipitates) are extremely stable at the intended service temperatures and have a major creep strengthening effect.

Boron has long been known to improve the creep strength of austenitic steels, even when added in very small quantities. However the mechanism of this beneficial effect has not yet been satisfactorily explained.

An improvement in long-term creep strength can be achieved by increasing the Mo-equivalent (Mo % + 0.5 W %) from 1.0 to 1.5 %. Fig.2 shows the creep strength for a temperature of 600°C and a duration of 30'000 hours for a number of 9 to 12 % Cr steels with different contents of W, Mo and B. The compositions of a number of steels investigated in the COST programme are also shown (2-4).

2.3 Development Concept

The approach adopted in the COST programme has been the following:

- Potential alloys for forgings and castings were identified after a critical review of the existing grades and steelmaking developments in Europe and elsewhere.
- Small trial melts (typically about 200 kg) were manufactured, heat treated and subjected to a standard mechanical testing programme, including measurements of strength, toughness, creep properties and resistance to long-term thermal instability.
- Trial components were manufactured from the steels showing the best combination of mechanical properties and were tested nondestructively and destructively with particular emphasis on uniformity of mechanical properties in large components and long-term creep and exposure testing.
- Components have now been manufactured and are successfully in service in power plant.

2.4 Forged Steels

The steel grades showing the best properties are designated as follows:

Steel Group B: Boron Grades. An earlier COST programme already identified the potential of steels containing about molybdenum and 100 ppm of B. Creep strength is improved and impact energy is not reduced at this level of boron.

Steel Group E: Tungsten/Molybdenum Grades. Three melts were selected with tungsten contents of 0.5 to 1.0%, while Mo level was retained at about 1 %.

Steel Group F: Molybdenum Grades. Test melts were selected with molybdenum level of about 1.5%, but no additions of W or B.

The standard investigation programme for the trial melts comprises:

- microstructural investigation
- tensile testing at ambient and elevated temperature
- impact testing at ambient temperature and transition temperature determination
- isothermal creep rupture tests using plain and notched samples
- isostress rupture tests at 100 MPa.

Creep strength data for the steels, plotted on the basis of the Larson-Miller parameter is shown in Fig.3. Creep testing times exceed 60 000 hours. The results show that the target creep strength values can be met. They lie considerably above those for the currently employed steel, standardised in the DIN standard SEW555.

Analysis of all results permitted the identification of those steels and heat treatments most promising for the production of full-scale rotor forgings. Fig.4 shows the sampling plan for one of the rotors. Using the following procedure it was possible to investigate two strength levels on a single rotor:

- Rotor was forged to shape and austenitised and provided with the first tempering treatment at 570°C
- Second tempering was performed at the lower temperature intended to produce a yield strength of > 700 MPa
- A radial trepanned core was removed from the rotor body to permit investigation of this strength level at any radial location in the rotor body
- The alternative second tempering treatment at the higher temperature was performed on the remaining part of the rotor body in order to attain a yield strength of > 600 MPa.

Results of the mechanical property tests for the lower yield strength condition of this trial rotor are also shown in Fig.4. Values are very uniform, showing little difference between rim and centre locations of the rotors. In addition in all rotors the amount of δ-ferrite is << 1 % and no significant segregation could be observed. Extensive mechanical testing of samples from different locations in the rotors has been performed, including Low Cycle Fatigue (LCF) lifetime, fracture toughness and resistance to subcritical crack propagation.

2.5 Cast Steels

A similar approach was employed in the development of cast steels. Based on the results of earlier testing and the evaluation of published information an attempt was made to identify compositions which would provide a further improvement in creep strength. On the basis of a cast version of the steel P91, an improvement was sought through modification of the heat treatment and through further alloying additions.

Accordingly melts either with or without 1 % tungsten were manufactured in the pre-evaluation programme. In each case the chromium content had been increased in comparison with P91 in order to improve the solubility of nitrogen, thereby increasing the potential for precipitation of MX (VN) which is believed to be a potent creep strengthening mechanism. Plates were cast and heat treated and subjected to a testing programme similar to that for the forged materials. Welding of cast steels is a requirement both for the repair of any minor cracks which may result during the casting process (manufacturing welds) and also for the joining of cast parts, such as valve bodies or turbine casings, to other components, such as steam pipes (design welds).

The longest creep tests of base material at 600 and 650°C are now approaching a duration of 55 000 hours. Fig.3 shows a Larson-Miller representation also of the creep rupture strength of tungsten-free and tungsten-containing material. The tungsten-containing cast version shows a slight superiority over the entire range of times and temperatures and lies even a little above the forged pipe steel P91.

The excellent behaviour of the tungsten-containing steel resulted in this composition being chosen for the component programme, in which a pilot valve chest of about 6 tons in weight was manufactured. Manufacture was not accompanied by any difficulties. Mechanical testing was carried out on the valve body and two manufacturing weld locations. Creep testing results confirm the results from the pre-evaluation programme.

In conclusion it can be stated that both cast and forged steels have been identified with a creep strength improvement of about 40°C in comparison with those previously employed. Representative forged and cast components have been manufactured and exhibit uniform properties. Components of these steels have now entered service in commercial steam power plant.

2.6 Steels for High-Temperature Pipework

The new piping steel T91/P91 (X 10 CrMoVNb 9 1) was developed at Oak Ridge National Laboratory in co-operation with ABB Combustion Engineering and is standardised in the USA (ASME Code). It provides an improvement compared with the European steel X20 CrMoV 12 1 and for this reason has been intensively investigated in the COST programme.
Long-term creep tests were performed to determine the properties required for design purposes. In the meantime testing times of more than 60 000 hours have been reached. Based on these results creep strength values for 100 000 hours have been estimated. They lie above those for X20 CrMoV 12 1. Higher creep strength values are already found at 550° C and with increasing temperature the percentage improvement compared with X20 CrMoV 12 1 rises significantly. Again there is a temperature advantage compared with X20 CrMoV 12 1 of almost 30°C (5). Extensive tests showed the ease of processing of T91/P91 by cold or hot bending. Manual metal arc, TIG and submerged arc welds of different thicknesses were manufactured. There are advantages as a result of the lower martensite hardness of the T91/P91. For example, cooling to room temperature is possible after welding.
Many thousands of tons of this steel are now in use in power plants throughout the world (6). In addition the TÜV has checked and approved its use in specific cases.

2.7 Metallographic Analysis

Metallographic investigations have been extensive (7). Specifically optical microscopy has been used to determine grain structure and uniformity in large components, quantitative transmission electron microscopy to correlate stability of microstructural features with mechanical properties and field ion microscopy, with which all elements present in the alloys can be precisely located and quantitatively analysed. Microstructural observations of the type, size and distribution of precipitates, along with their evolution at high temperature as a function of time and applied stress, have permitted models to be developed to describe and predict the volume fractions of precipitates both after heat treatment and also extended periods of service. This work is leading to the development of further generations of improved alloys.
The metallurgy of this class of steels is complex, the long-term strength being dependent on the precipitate distribution generated by heat treatment and its stability in the long term, along with further precipitation reactions which occur successively during extended periods at elevated temperatures. An "ideal" microstructure in this class of steels could be considered to comprise a tempered martensite with a small lath size and a high density of dislocations stabilised by a large number of finely distributed hard particles.

Qualitatively it is known that during long-term thermal exposure the following effects can be observed:
- Laves phase containing a high proportion of W and Mo is inhomogeneously nucleated and coarsens thereby reducing the extent of solid-solution strengthening.
- $M_{23}C_6$ particles are coarsened so that their effect in maintaining the high dislocation density and hence the high creep strength is partially lost.
- MX particles remain essentially stable.

Metallurgical understanding of the processes controlling microstructural stability can be obtained by quantitative analysis and modelling of the diffusion-controlled processes.

Computer software packages, such as Thermocalc (8), have recently become available for the prediction of the phases which will be present at a certain temperature in steels such as those presently under discussion. The programme is able to predict from the chemical composition of the steel alone the amount and composition of phases such as δ-ferrite, $M_{23}C_6$, Laves Phase and MX under equilibrium conditions. Fig.5 shows a typical example for the steel Nf616 containing about 2 % W (9). Each of these phases has been identified in the steels under consideration.

However kinetic processes of nucleation and growth of the precipitates are of major importance and long times are required before equilibrium conditions are reached (9). A higher temperature of exposure leads to a more rapid precipitation of Laves phase and a more rapid attainment of the equilibrium condition, but the total amount precipitated in the equilibrium condition is correspondingly lower. At a typical service temperature of 600°C the equilibrium condition may only be reached at about 20 000 hours. At temperatures above 700°C the phase is thermodynamically unstable.

One specific observation concerns the apparent effect of boron. In the steel B it is seen that there is a larger number of the $M_{23}C_6$ precipitates and their coarsening rate appears to be lower than in boron-free steels. However transmission electron microscopy (TEM) was not able to reveal the mechanism by which boron was producing this beneficial effect. Lundin (9) used the technique of atom probe field ion microscopy (APFIM) to examine materials with and without boron additions on the atomic scale. The technique has provided the following results:

$M_{23}C_6$

- Nucleation of $M_{23}C_6$ precipitates is faster and their coarsening rates may be slower than in the absence of boron
- In the as-tempered condition about half of the boron is present within these carbides
- A change in composition of the $M_{23}C_6$ is seen during creep - the composition tends from the equilibrium value at the tempering temperature towards the equilibrium value appropriate to the creep testing temperature. The process is limited by the corresponding diffusion rates.

MX

- No boron has been observed within the MX particles, but in the presence of boron their morphology changes from discs to needles.

The possibility of a "latent" creep strengthening effect as a result of precipitation of MX particles on dislocations is proposed. This may be a type of dynamic regenerative precipitation.

3. DESIGN ASPECTS

Use is made exclusively of proven design principles.

3.1 High Pressure Turbine

The main design features of the HP turbine section for the high temperature process are shown in Fig.6. The proven design principles of this turbine series can be applied without any modification. Only the inner casing, valve bodies and middle rotor section are manufactured from the new 10 % chromium steels.

Since the steam expands and cools in the direction of the rotor ends, a low-alloyed 1 % CrMoV steel is used in these regions. Because of the good running properties of this steel, there is no need for overlay welding of the rotor journals, which are stressed by bending and torsion. The weld between the high-alloyed and low-alloyed steel is made using conventional methods and filler metal. Operating temperatures and loads do not exceed the usual present-day values.

Additional advantages of the welded rotor are:

- Simplicity and accuracy of non-destructive testing
- Wide selection of forging suppliers
- Uniformity of mechanical properties and high toughness value

An advantage for medium-load operation results from the rotationally symmetrical inner casing constructed with a shrink-ring connection. By avoiding non-uniform wall thickness, no inadmissible high thermal stresses and hence plastic deformations occur. Situated in the HP exhaust, the shrink rings are always cooler than the inner casing and the necessary shrinkage force is maintained under all operating conditions. Thus, the load change and start-up gradients of this design do not differ from those for turbines operating with lower live-steam temperatures.

The diffusers which pass the steam from the inlet valves to the HP turbine are of the free-expanding type and equipped with piston ring sealing elements in the inner casing. They are also located in the cooler HP exhaust steam. The advantage is that no part of the outer casing is exposed to the live-steam temperature. This means that the HP extraction required for the process can be achieved with very little effort. Such an extraction concept has already been employed in many steam power plants.

3.2 Intermediate Pressure Turbine

The IP turbine section operates with an inlet temperature of 600°C. Despite the high reheat temperature, the entire design concept can be adopted without change.

Fig.7 shows the material changes in the IP turbine section. Only the inner casing, the valve bodies and the middle rotor section are made of the new 10 % chromium steels. The IP rotors of ABB turbines are designed with smaller diameters than those of other manufacturers so that there is centrifugal forces can be sustained without any requirement for steam cooling of any components. Steam cooling would result in a loss of operating efficiency. The IP steam inlet from the laterally positioned valve casings is made directly into the inner casing, just as for the HP turbine. The control and stop valves upstream of the IP turbine can be connected with the IP casing using short pipe sections of the steel P91. This steel has been assimilated in the ASME Standard and should also be included in DIN.

The halves of the IP casing are held together by bolts which are located in the exhaust steam from the IP turbine. They can be manufactured from the usual steels as their temperature does not exceed 500°C.

In both the HP and IP turbine the first rows of rotating blades are made of the austenitic steel X12CrNiWTi 16 13, as is already the case for all supercritical turbines operating at above 540°C. Even at 600°C this is still well within its operating limits.

The LP turbines operate under conventional steam conditions.

4. FURTHER DEVELOPMENT

Steels which have so far been manufactured and tested in the COST programme have properties which permit their use in steam power plant at temperatures up to 600°C. However there is a considerable driving force to increase temperatures of operation even further in order to obtain further improvements in operating efficiency [1]. A net operating efficiency of about 50 % can be attained with steam inlet temperatures of about 650°C. Based on the development activities initiated in the present phase of the COST programme it is anticipated that such plant could be constructed and commissioned by the year 2005. For this purpose use is being made of the information and understanding gained concerning microstructural stability at elevated temperatures along with modelling activities which describe not only the equilibrium phases, but also the kinetic processes governing microstructural evolution. This constitutes a complex process of continuous nucleation, growth, coarsening and redissolution of different phases.

A further major step could then be made through the use of **nickel based alloys** for the high-temperature components in the steam cycle. In order to justify the use of these considerably more expensive materials a large improvement in operating efficiency will be required, and for this reason inlet temperatures are expected to rise to 700°C or more [1], as shown in Fig.8. At the same time it will be necessary to minimise the amount of such materials employed and to develop transition joints between these and lower alloyed materials. A concept for the realisation of such a high-efficiency steam power plant is under development [1,11]. In this design major large components such as rotors, blades, casings and main steam pipes will be exposed to temperatures of up to 720°C. Use can be made of experience obtained from the design of gas turbines. However the components will be of larger size and particular emphasis must be placed on the question of up-scaling. For construction of plant with a generating capacity of 800 MW, rotors will be required with diameters of typically one meter and weights in excess of 20 tons. Castings will have large outer diameters (about two meters). There is as yet no experience with the manufacture of such large parts in nickel-based alloys. Attention will have to be paid to the subjects of segregation in large cast blocks, forging capability of available presses, uniformity of mechanical properties in large components and non-destructive testing capability.

Alternatively it may be appropriate to employ manufacturing techniques, such as powder metallurgy, in order to avoid segregation problems, or welding, as a means of building up large components from smaller forgings or castings.

5. CONCLUSIONS

The research and development activities described here have resulted in the validation and application of a number of 9 to 12 % Cr steels with considerably improved high-temperature properties. They are appropriate for use at temperatures up to 600°C. Detailed microscopic investigations have led to an improved understanding of the mechanisms of creep strengthening and microstructural stability, on the basis of which further efforts are being made to raise the upper limiting temperature for a further range of steels. These are currently being manufactured and tested.

It is not anticipated that major design changes will have to be made to permit steam inlet temperatures to be raised to 650°C, thereby increasing operating efficiency to about 50 %.

A further major increase in operating temperature of steam turbines to 700°C or more will only be possible through the use of nickel-based alloys. Whereas advantage can be taken of experience from the gas turbine field, larger component sizes will pose new questions. Efficiencies may reach 55%.

6. ACKNOWLEDGEMENTS

The author is grateful to his colleagues and partners in the COST programme for their contributions to some parts of the work reported here and for many helpful discussions during the course of the work. Thanks are also extended to the COST Management Committee for their guidance of the programme and to the national funding bodies for their financial support of the individual projects.

7. REFERENCES

[1] S.Kjaer, "Anforderungen an das Kraftwerk 2000/2015", Conference on Coal-Fired Power Plant in 2000/2015, 30th and 31st March 1995, Dresden, Germany, p.41

[2] C.Berger, R.B.Scarlin, K.H.Mayer, D.V.Thornton and S.M.Beech, "Steam Turbine Materials: Forgings", COST Conference on High Temperature Materials for Power Engineering 1994, Liège, Belgium, Oct. 3 - 6 1994, p.47

[3] R.B.Scarlin, C.Berger, K.H.Mayer, D.V.Thornton and S.M.Beech, "Steam Turbine Materials: Castings", ibid, p.73

[4] C.J.Franklin and C.Henry, "Material Developments and Requirements for Advanced Boilers", ibid, p.89

[5] C.Berger, W.Bendick, K.H.Meyer and R.B.Scarlin, "Neue ferritische martensitische Stähle für Temperaturen über 550°C für Rohrleitungen und Dampfturbinen". VGB-Jahrbuch der Dampferzeugertechnik (1992).

[6] J.Orr and D.Burton, "Development, Current and Future Use of Steel 91", ECSC Information Day, November 5th, 1992, VdEh, Düsseldorf.

[7] R.W.Vanstone, "Microstructure and Creep Mechanisms in Advanced 9-12 % Cr Creep Resisting Steels - A Collaborative Investigation in COST 501/3 WP11", COST Conference on High Temperature Materials for Power Engineering 1994, Liège, Belgium, Oct. 3 - 6 1994, p.465

[8] B.Sundman, B.Jansson and J.O.Andersson, Calphad, 9, 2, 1985, 153 - 190

[9] J.Hald, "Materials Comparisons between Nf616, HCM12A and TB12M - III: Microstructural Stability and Ageing", EPRI / National Power Conference, 11th May 1995, London, p.152

[10] L.Lundin and H.O.Andren, "Atom Probe Investigation of a Creep-Resistant 12 % Chromium Steel", in Surface Science 266 (1992) p.397

[11] C.Berger, R.B.Scarlin, K.H.Mayer, "Werkstofftechnische Entwicklungsaufgaben für Dampfturbinen mit Eintrittstemperaturen von max. 700°C", ibid, p. 239

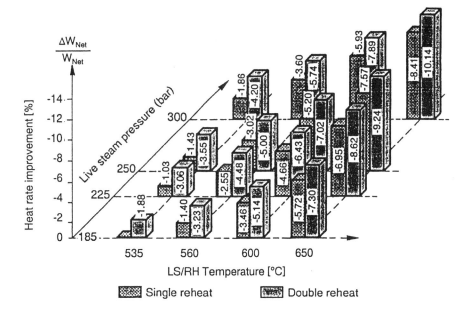

Fig.1: Process Improvements for Single and Double Reheat

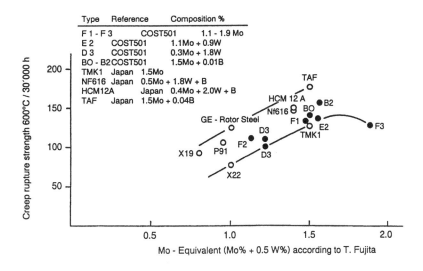

Fig.2: 30 000 hour Creep Rupture Strength of 9 to 12 % Cr steels as a function of Mo-Equivalent

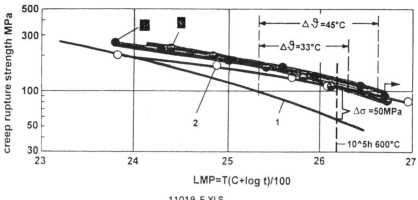

Forged Steels	C	Cr	Mo	W	V	Nb	N	B	Rp0.2 RT
1 X21CrMoV121	0.23	12	1.0	-	0.30	-	-	-	min.600 MPa
2 X12CrMoVNbN101	0.12	10	1.5	-	0.20	0.06	0.05		- 600 MPa
3 X12CrMoWVNbN101	0.12	10	1.0	1.0	0.20	0.06	0.05		- 700 MPa
4 X18CrMoVNbB91	0.18	9.3	1.5	-	0.27	0.06		0.01	- 650 MPa

(a)

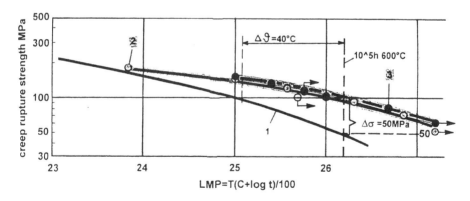

Cast Steels	C	Cr	Mo	W	V	Nb	N	Rp0.2 RT
1 G-X22CrMoV121	0.22	11	1.1	-	0.25	-	-	min.590 MPa
2 G-X12CrMoVNbN101	0.12	10	1.0	-	0.22	0.07	0.05	582 MPa
3 G-X12CrMoWVNbN1	0.12	10	1.0	1.0	0.22	0.07	0.05	592 MPa

(b)

Fig.3: Creep Behaviour of Improved Forged and Cast Steels.

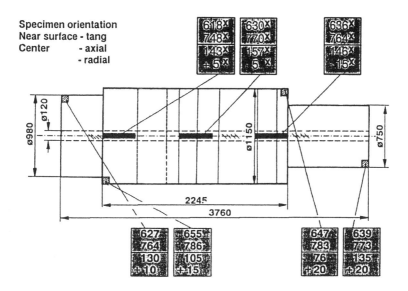

Fig.4: Mechanical Properties at different Locations in Trial Rotor E (Yield Strength > 600 MPa).

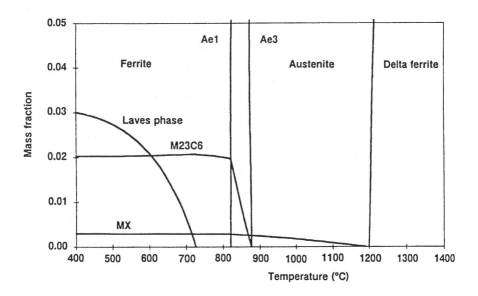

Fig.5: Equilibrium Phases in the Steel Nf616 (calculated by Thermocalc)

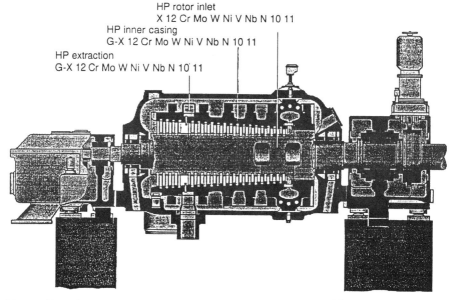

Fig.6: Longitudinal Section through the High-Pressure Turbine.

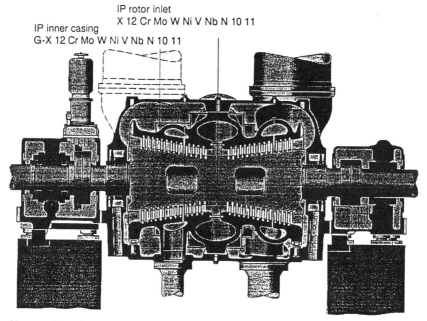

Fig.7: Longitudinal Section through the Intermediate-Pressure Turbine.

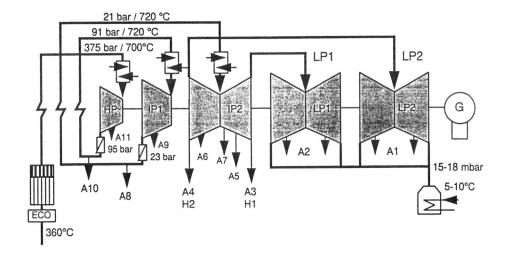

Fig.8: Coal-Fired Steam Power Plant with Ultra-High Efficiency.

C522/032/97

Advanced steam turbines for modern power plants

Dr Ing **H OEYNHAUSEN**, **A DROSDZIOK** Dipl Ing, and **M DECKERS** Dipl Ing, PhD, VDIGermany
Steam Turbine Development, Siemens Power Generation (KWU), Mülheim an der Ruhr, Germany

INTRODUCTION

One of the primary tasks associated with the development of modern power plant facilities is to improve the economy of operation while, at the same time, reducing the environmental pollution. Further goals are high reliability and safety of operation, high availability, short construction and commissioning times, a long service life, as well as low maintenance costs. In order to meet these numerous requirements, all components of the power plant must undergo a comprehensive optimization process in the overall context of the plant as a whole. The outcome is considered to be a new generation of steam turbine power plants. Naturally, the steam turbine is one of the key-components in the power plant and, in the light of this, Siemens has recently designed a new turbine series to meet the challenges of the future.

The overall cycle efficiency of the plant is very strongly related to the efficiency of the steam turbine. As a consequence, efforts to enhance the efficiency are focussed increasingly on the reduction of energy losses in all components of the steam flow path. Considerable progress has been made in recent years particularly by improving the turbine blades. As a result of that, all turbine components can be equipped with highly efficient, three-dimensional blading these days. At the same time, however, attention is also being increasingly drawn to optimise all the other flow-path components, e.g. casings, pipes and valves.

OVERVIEW OF SIEMENS STEAM TURBINE SERIES

Siemens new turbine series for fossil-fired steam power plants with single or double reheat and for combined-cycle power plants cover an output range from about 100 to 1200MW for both the 50Hz and the 60Hz market. The series builds on five basic turbine modules, **Figure 1**. There is one module designed to operate in each of the three pressure ranges, i.e. the high pressure (HP), intermediate pressure (IP) and low pressure (LP) range. In addition to that, high and intermediate pressure turbine modules as well as intermediate and low pressure turbine modules can be combined in a single casing to form HP/IP turbine modules and IP/LP turbine modules, respectively. For each module there is a standard size, from which several size versions are derived to cover the entire range of output ratings with allowance for different steam parameters and turbine speeds. Various modules can be combined to meet all further requirements resulting from the plant process. The five basic modules (HP, IP, LP, HP/IP and IP/LP) in their various versions are complemented by further standardised elements such as valves, bearings, and cross-over pipes to form the turbine unit. **Figure 2** displays a longitudinal cross-section of a typical large advanced power plant unit. In the present case, the unit is made up from a combination of HP, IP and LP turbine modules featuring separate HP and IP turbines and three double-flow LP turbines. The great advantage of this design philosophy is that the individual modules are replicated in large numbers. This not only ensures a high rate of operating experience feedback for enhanced reliability and high availability but it also facilitates achievement of an economic solution for any given application.

In order to be able to achieve the high efficiencies required for the new generation of power plants, the series has been designed for main steam pressures and temperatures of 300bar and 600°C, respectively, and for reheat temperatures of 600°C. To accomodate these high pressures and temperatures, both the HP and IP turbine modules feature special design characteristics and use of 10% chromium steel in their admission sections.

The use of advanced steam conditions has been standard engineering practice for many years now as is demonstrated by **Table 1**. In the period between 1951 and 1991 Siemens Power Generation (KWU) provided some 44 turbines with either main steam pressures above 245bar or main steam temperatures above 550°C (with a maximum main steam temperature of 640°C). Orders have been placed for seven steam turbine units in a rating range of 395MW to 1000MW.

In Germany, further developments were achieved which are characterised by the following features:

- High-temperature cycle with atmospheric coal firing, single reheat and steam conditions of up to approx. 270bar / 580°C / 600°C
- Specified net efficiency of 45% and above
- Nine-stage feedwater heater train
- Optimized cold end
- Heat extraction for process and district heating
- Compliance with stringent German enviromental protection regulations for power plants

Number of Orders	Power [MW]	Main Steam Conditions [bar/°C/°C]	Start of Operation
44	7 - 700	> 245 / or > 550	1951 - 1991
1	395	263 / 540 / 565	1997
2*	874	264 / 542 / 560	1997
1*	750	250 / 575 / 600	1999
2*	910	260 / 540 / 580	1999 - 2001
1*	1000	250 / 580 / 600	1999

* use of 10% chromium steel

Table 1: List of Orders with Advanced Steam Conditions

The following comments on the design features of the respective turbine modules refer to the classic HP/IP/LP turbine series for high output ratings.

High Pressure Turbine

The single flow HP turbine module continues in the tradition of the well-proven barrel-type design, **Figure 3**. This design eliminates the need for flanges along the sides of the outer casing and brings about a particularly compact, circumferentially symmetrical geometry. It ensures that no casing deformation occurs during operation and, hence, there is no risk of non-uniform changes in the radial clearances. Even very advanced steam conditions and changes in load and temperature will cause moderate thermal stresses only. It should therefore be stressed that the advantage of this design compared to turbines with an axial joint increases as steam conditions become more elevated. At such elevated steam conditions, boilers are operated in the sliding-pressure mode and HP turbines equipped with full-arc steam admission are employed. The steam enters the blading through the two side-mounted combined stop and control valves and the inlet section, **Figure 3**. Making use of a weak reaction stage as the first blading stage reduces the rotor temperatures in this region. At the same time, the clearance-free seal at the stationary blade row enhances the efficiency of this stage.

Intermediate Pressure Turbine

The single or double flow IP turbine module is designed as a two-shell, horizontally-split cylinder, **Figure 4**. The steam enters the turbine via the two combined stop and control valves located below the horizontal joint. It is then fed through the blading, exhaust casing and cross-over pipe located on top of the turbine to the low pressure section. All extraction points are located on the bottom half of the casing. A weak reaction stage in conjunction with a rotor thermal barrier shield is used in the IP turbine module to enhance the efficiency. In case that reductions of the stresses due to high steam admission temperatures or oversize dimensions are required, the shaft thermal barriershield can be equipped with a vortex cooling device, **Figure 5**. Within this cooling device, steam expands through a number of tangential bores into the cavity between the shield and the rotor and, consequently, its temperature drops. Since both the rotor and the steam rotate in the same direction at roughly the same speed, the expanding steam forms a cooling blanket around the shaft.

Low Pressure Turbine

The LP turbine modules for the new series are essentially based upon the well-proven and tested predecessor LP turbine design with the inner casing supported directly alongside the bearings, **Figure 6**. A push rod connection from the outer casing of the IP turbine enables the inner casing of the LP turbine to follow shaft expansion towards the generator. This design facilitates the reduction of the axial clearances in the second and third LP turbines. A simple lever mechanism is provided between the push rods of the second and third casing to increase the axial movement. In doing so, the axial clearances in the third LP turbine can be further reduced.

IMPROVING THE EFFICIENCY OF THE STEAM TURBINES

The availability of reliable three-dimensional fluid flow computation programs enables the optimisation of steam turbine blades with respect to friction, leakage flow as well as secondary flow effects. At the same time, however, efforts are also increasingly being made to optimise the non-bladed flow channels such as the valves, pipework, admission and exhaust casings from the point of view of fluid mechanics.

The present section describes a number of numerical and experimental studies that were carried out aimed at optimising the fluid mechanic aspects of steam turbine components. Design criteria for modern, three-dimensional, high-performance turbine blades are discussed and improved admission and exhaust section geometries of turbine modules are presented. Results of numerical investigations making use of a three-dimensional, viscous computation procedure will be presented, discussed, and compared with experimental evidence.

Optimisations of Steam Turbine Blades

High Pressure and Intermediate Pressure Blades

Figure 7 shows the relative flow losses at various locations in the HP and IP turbines, respectively. Generally speaking, the blade profile loss (the loss due to boundary layer growth along the blade surface and due to dissipation in the blade wake) can be seen to be largest single source of loss in the whole turbine. This clearly indicates that the overall loss can be most easily reduced by reducing the profile loss throughout the entire turbine. The so-called secondary loss (the loss due to viscous effects in the endwall boundary layers at the hub and casing) is of significant magnitude for those turbine stages that are characterised by a low aspect ratio (the ratio of blade span to blade chord), i.e. the front stages of the HP and IP turbine. It is therefore desired to reduce the secondary loss in these stages. Leakage losses are relatively high in the admission sections to the HP and IP turbines. Leakage-free admission segment designs, e.g. a one-piece admission segment, can therefore result in considerable efficiency gains in these parts of the turbine.

Figure 8 shows the development history of cylindrical blade profiles over the past decades; beginning with the so-called T2-profile in the 1970's, via the T4-profile in the 1980's and the improved TX-profile in the 1990's. The recently released TX-profile provides an improved overall stage efficiency compared to the T4-profile while, at the same time, being less prone to deposits on the suction surface. The earlier T4-profile was suitable for a wide range of applications due to its very flat optimum efficiency curve whilst the TX-profile yields advantages for part-load operation over a predefined load range. This is in accordance with current operating practice for large steam turbines.

The next step in blade development for enhanced efficiency is the so-called 3DS blade. This new blade type was specially designed for use in the front stages of HP and IP turbines. The blade geometry is characterised by a three-dimensional design aimed at reducing the secondary losses at the hub and casing of the blade. Detailed experimental investigations performed on a 4-stage test turbine show that an efficiency improvement of up to 2% can be achieved with these 3DS blades in comparison to the conventional cylindrical blades.

The final stages of HP and IP turbines as well as the front stages of LP turbines are usually equipped with twisted blades featuring integral shrouds, **Figure 9**. Considering the relatively high aspect ratio, 3DS correction at the root and tip of the blade yields no significant advantage because the influence of the secondary flow decreases with increasing blade span.

Low Pressure Blading

As mentioned above, the front blade rows of the LP turbine are designed as twisted blades with shrouds. The final stage in the exhaust region is however quite different: The rotor blades are of a free-standing type without shrouding or damping elements whilst the hollow stator blades can be equipped with suction surface slots for drainage or hot steam heating, respectively. Both measures can be introduced to reduce wetness losses and blade erosion.

Free-standing blades were first used in the 1950's and at that time represented a milestone in the development of large exhaust cross-sections for LP turbines. In the mid-80's, sophisticated computer programs were first used for the three-dimensional design of turbine stages. One outstanding result of the three-dimensional flow analysis was the demonstration that curved stationary blades enable an optimal flow distribution along the entire blade span (height), **Figure 10**. The velocity distribution at the blade tips was also optimised by the computation of the fully three-dimensional flow. The velocity at the blade tip is supersonic and so a "backward-curved" blade profile is used to avoid high local supersonic speeds. In comparison to earlier designs with simple nearly "flat plate profiles" at the tip, local velocity peaks can be reduced by up to 20%.

The last stages of the LP turbine are designed with drains along the circumference. One highly effective way of avoiding droplet impact erosion in the final stage is to use slotted hollow stator blades. The water film on the blade surface can then be drawn off through slots. A new and even more effective method, is to heat the final stationary blade row, **Figure 11**. For this purpose, steam from an extraction point is introduced into the hollow stator blades. The heating of the blades causes the moisture film on the blade surface to be evaporated. As a consequence, water droplets can be avoided which otherwise could be entrained by the steam and, in turn, could damage the leading edges of the last rotor.

Optimization of the Steam Turbine Admission and Exhaust Geometries

In the context of efforts to improve the steam turbine efficiency, the entire steam flow path from the main steam valves to the condenser must be considered. Within this process of development, comprehensive numerical and experimental investigations were performed on scaled models as well as on installed turbines in the power plant. The measurements served also to derive realistic boundary conditions for the computations and to validate the computational methods. Comparisons between numerical calculation and experimental evidence confirmed that modern computation procedures are by all means able to compute complex, three-dimensional flows with sufficient accuracy provided that realistic boundary conditions were specified. This is demonstrated by examples in the following.

Intermediate Pressure Turbine Exhaust Casing

In parallel with the computations, experimental fluid flow studies were performed on scaled models of a typical IP turbine module. The aim of the investigation was to study the complex three-dimensional flow field that occurs in the exhaust casing and, particularly, to address the question as to how it is influenced by the diffusor geometry. In order to obtain realistic flow conditions at inlet to the computational domain, the flow resistance introduced by the blade rows was also simulated by means of honeycombs and screens. The measurements performed served essentially to determine the global (total) pressure loss.

The flow computations were carried out using the commercial CFD package TASCflow[1]. Since the flow in the exhaust casing was experimentally found to be symmetric about the two main axes, it was possible to reduce the computational domain to one quarter of the overall casing. The geometry and computational mesh is presented in **Figure 12**. The simulation covered the flow from the exit of the blading through the diffusor and annulus to the exhaust flange.

The computed velocity distribution in the symmetry planes is shown in **Figure 13**. The results show that the flow is not able to follow the sharp changes of curvature pertinent in the diffuser. As a consequence, the flow separates and re-circulation zones are created. This gives rise to vortices which are accompanied by relatively high energy losses. **Figure 14** demonstrates that these vortices exist in the entire exhaust casing and that they cause swirl velocity components to be generated in the exhaust pipe.

On the basis of these results, studies with modified diffuser geometries were carried out aimed at improving the guidance of the steam flow. The findings indicate that optimizing the component geometry can considerably reduce flow separation in the diffuser and vortex formation in the casing. Hence, the pressure losses in the IP turbine exhaust casing can be minimised.

Low Pressure Turbine Exhaust Casing

The following presents the results of numerical computations of the flow field in the exhaust casing of a low pressure turbine. These calculations were again performed using the CFD code TASCflow.

In the symmetrical low pressure turbine, the steam flows from the middle of the turbine in both axial directions. Downstream of the final stage of blades the flow passes through the diffusor to enter the exhaust casing. This is the domain under consideration. **Figure 15** shows a typical configuration that was investigated. At the inlet plane to the computational domain the flow still has a swirl component. Because of this swirl and of the asymmetric location of two extraction lines that run downwards through the exhaust casing,

---

[1] TASCflow User Documentation (1995), Advanced Scientific Computing Ltd., Waterloo, Ontario, Canada.

the computation has to be carried out for the entire exhaust casing. For this reason too the lines from the internal extraction points to the low pressure feedwater heaters were incorporated into the computational model.

Figure 16 shows the results of the computation by means of streamlines. It is evident from these figures that the flow field in the exhaust casing is characterised by pronounced vortices and swirl components being superimposed on the main flow direction and hence energy losses are generated. However, flow guidance can be improved by making use of shields, baffles, or similar elements.

SUMMARY

The modular principle of Siemens' new turbine series ensures optimum matching of the diverse requirements of modern power plants. A high level of replication of the individual components means that comprehensive operating experience feedback is available for further enhancing the reliability and safe operation of the turbines.

The new generation of steam turbines for power plants marks a milestone in the evolution of turbine blading. The availability of reliable three-dimensional viscous CFD codes enables the computation of the complex flow through turbine blading. The introduction of these methods allows the design engineer to account for viscous, secondary flow and tip leakage effects very early in the design process. As a result, stage efficiency improvements of up to 2% are possible. Since still higher efficiencies are increasingly being required, it is strongly recommended to investigate also the non-bladed components (e.g. casings, valves, pipes, etc.) along the steam flow path for potential improvements. Component geometry optimisations as well as the addition of flow-guiding devices can considerably help to reduce flow separation as well as vortex formation and thus energy losses can be minimized.

FIGURES

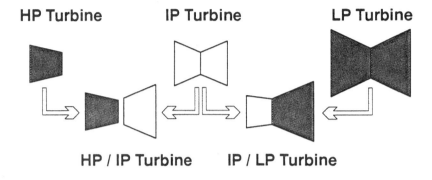

Figure 1: Basic Modules of Turbine Series

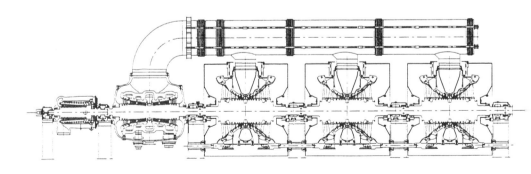

Figure 2: Longitudinal Cross Section of Typical Advanced Power Plant Unit

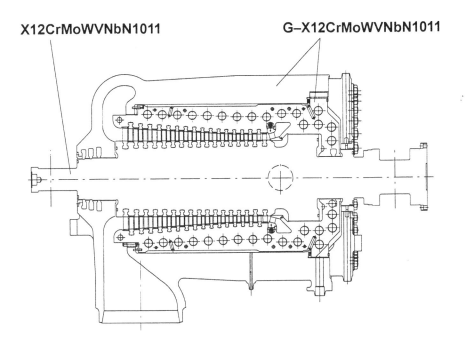

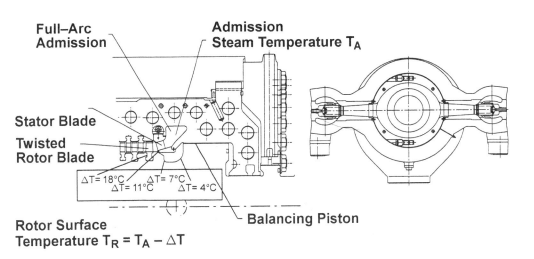

Figure 3: HP Turbine Design with Full-Arc Admission

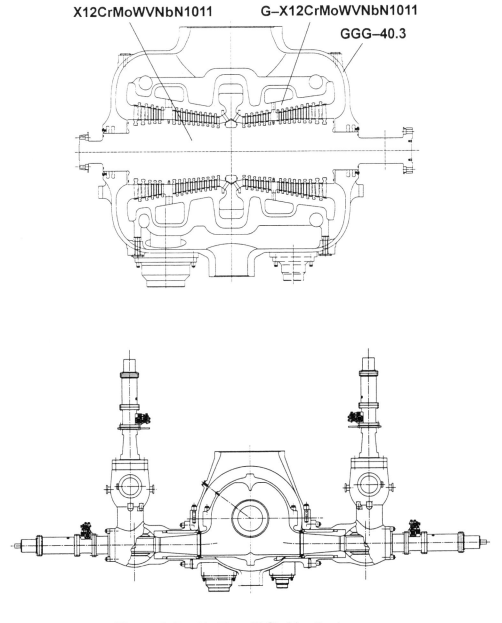

Figure 4: Double-Flow IP Turbine Design

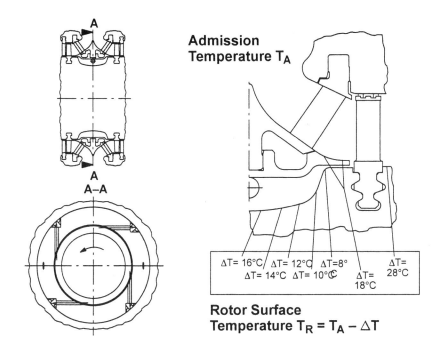

Figure 5: IP Turbine Design with Shaft Steam Shield and Vortex Cooling

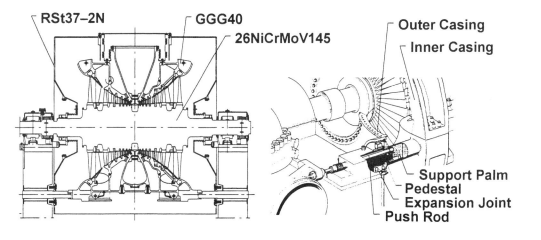

Figure 6: LP Turbine Design with Push Rod Connection

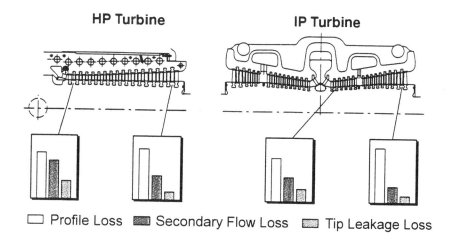

Figure 7: Relative Flow Losses in HP and IP Turbine

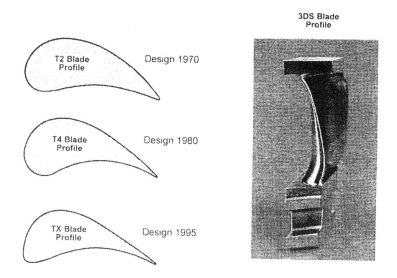

Figure 8: Development History of Steam Turbine Blades

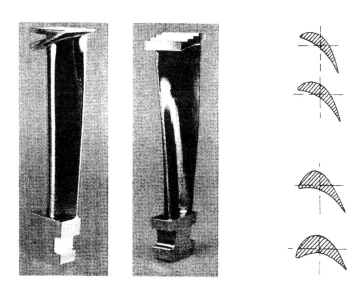

Figure 9: Twisted Turbine Blades with Intregral Shrouds

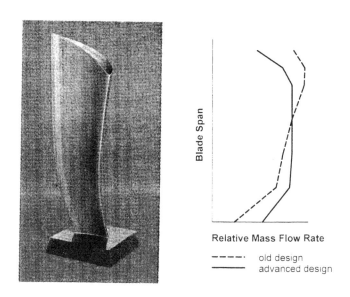

Figure 10: Comparison of Stator Blades in Final Stage

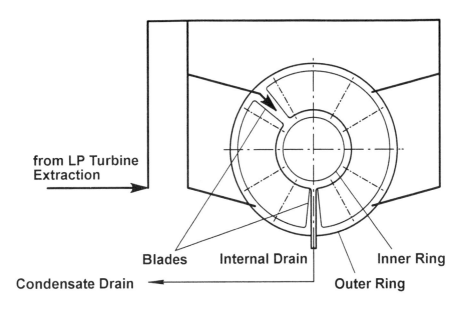

Figure 11: Heated Hollow Stator Blade Diaphragm

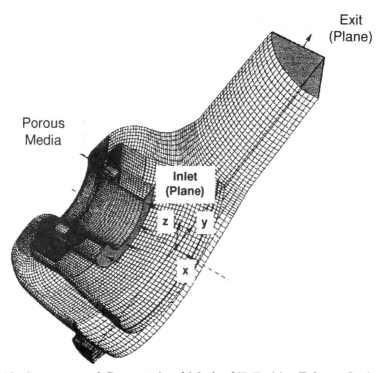

Figure 12: Geometry and Computational Mesh of IP Turbine Exhaust Casing

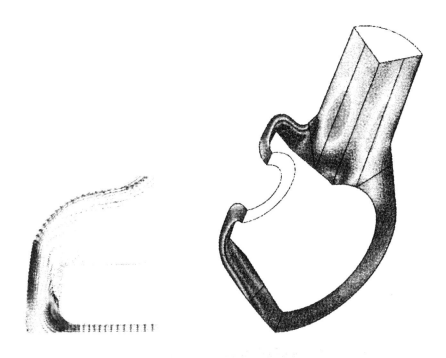

Figure 13: Velocity Distribution in Symmetry Planes of IP Turbine Exhaust Casing

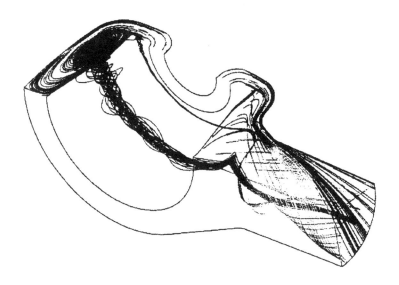

Figure 14: Vortex Generation in IP Turbine Exhaust Casing

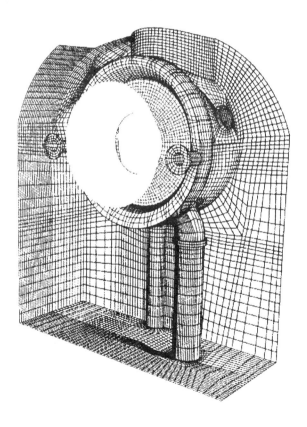

Figure 15: Computational Mesh of LP Turbine Exhaust Casing

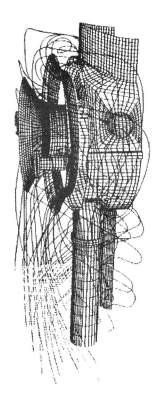

Figure 16: Streamlines in LP Turbine Exhaust Casing

Materials for Steam Turbines

C522/007/97

New materials for advanced steam turbines

R W VANSTONE MA, CEng, MIM and **D V THORNTON** CEng, FIM
GEC ALSTHOM Steam Turbine Group, Rugby, UK

SYNOPSIS

Since the mid 1970s major efforts have been made to improve the creep strength of 9-12%Cr steels. The first steel to emerge from this work, modified 9%Cr1Mo, was originally developed in the USA but has now been applied in pipework form across the world. This steel has also been applied in casting, plate and forged form. In Japan and Europe similar steels were developed including steels for application to rotor forgings. The availability of these steels has enabled the construction of steam plant with operating temperatures of up to 600°C.

New alloys are under development or just becoming available for application and will enable plant with even higher operating temperatures. Ultimately it may be possible for operating temperatures approaching 650°C to be achieved using 9-12%Cr steels.

Beyond 650°C it will be necessary to turn to a new family of steam turbine materials. Austenitic steels offer only marginal advantage in creep strength over the strongest 9-12%Cr steels and suffer from poor resistance to thermal cycling. Nickel based alloys offer much higher creep strength and have been applied for many years in gas turbine technology. A European initiative is in progress to adapt these alloys to enable the construction of steam plant operating at temperatures above 700°C.

1. INTRODUCTION

GEC ALSTHOM has been and continues to be very active in the development of new materials for advanced steam turbines, often in national collaborative programmes, in European collaboration through frameworks such as COST, or globally through EPRI collaboration, but also through private work in our own laboratories. In order to develop new turbine materials to the point of application it is necessary to generate long term creep data, to have conducted production trials on representative components and, where appropriate, to have developed necessary welding procedures. The costs of such development mean that it is important to identify the most promising alloys from the range of candidates as early as possible and to concentrate subsequent development on the smallest possible number of alloys suitable for the full range of turbine components.

2. BACKGROUND

2.1 Materials applied to conventional steam turbines

Until very recently there had been no major changes in the materials applied to the highest temperature components in steam turbines for over 30 years. Since the late 1950s low alloy CrMoV materials had been used for rotor forgings and castings while 12%CrMoVNb steels had been used for blading. During this period developments had focused on the understanding and optimisation of the properties of these materials and on increasing the reliability of components and fabrications manufactured in these steels (1,2). In the UK these activities took place through collaboration between CEGB, the turbinemakers and the steelmakers, long term data being gathered collaboratively and also through private development.

After extensive metallographic and creep studies led by GEC ALSTHOM with the collaboration of steelmakers, means of ensuring optimum microstructure in low alloy CrMoV for application to rotor forgings were developed. Microstructure was controlled through careful specification of chemical composition and heat treatment to give optimum properties, especially high creep strength (3). Maximum creep strength is obtained when the microstructure is upper bainitic with a dense dispersion of vanadium carbide particles (Figure 1). In order to achieve such a microstructure in thick sections it is necessary to control hardenability through appropriate additions of Cr, Mo, Mn and Ni. If the levels of these elements are too high a lower bainitic structure results and if they are too low ferrite is likely in the centre of the forging. In either case reduced creep strength may result. An optimum dispersion of carbide particles is obtained by control of V:C and Mo:V ratios. Large grain size can result in poor ductility and enhanced tendency to embrittlement so grain size is controlled through selection of the appropriate solution treatment temperature.

In CrMoV fabrications problems of reheat cracking were widely encountered in the coarse grained HAZ. Heat treatment and stress relief parameters, V:C ratios, the presence of elements like Sn, As and Sb and heat input during welding were all found to be significant factors in this phenomenon (4). Today the problem has been virtually eliminated through improved steelmaking and heat treatment practices, the application of low heat input welding procedures and the adoption of materials with V:C ratios giving reduced susceptibility.

The 12%CrMoVNb materials used for high temperature blading were derived from materials developed for jet engine applications, such as the FV448 alloy developed in the UK. The high alloy content of these steels gives them high hardenability so that on cooling after solution treatment martensite is formed even in thick section and at relatively low cooling rates. This martensitic matrix contains a dense network of dislocations which is stabilised by a dispersion of carbides formed during tempering (Figure 2). Dislocation movement is inhibited by interaction with the particle dispersion and with the high density of other dislocations, resulting in high creep strength. These steels contain high levels of C which combines with large amounts of carbide forming elements like Mo and V but especially Nb. Although these steels were stronger than the low alloy CrMoV steels used for rotors and castings they were not generally considered appropriate for such components, being susceptible to segregation and having poor forgeability, poor weldability and limited fracture toughness. Even their advantage in creep strength is limited; while these steels have very high creep strength for durations of 10 000 hours, appropriate to their jet engine application, at temperatures above about 550°C their creep strength falls rapidly as durations are extended towards the 100 000 hours which is the typical basis of steam turbine design (5). This reduction in creep strength occurs as a result of microstructural instability; the carbide dispersion becomes much coarser after long durations and is then less effective in inhibiting the movement of dislocations. Despite the potential segregation problems, certain turbine manufacturers considered their advantage in creep strength sufficient to apply slightly modified versions of these alloys to rotor forgings.

Slightly simpler 12%CrMoV alloys have been more widely applied, particularly in pipework form. However the high carbon content of these alloys and the influence of this in depressing martensite transformation temperatures meant that careful control of temperature during and after welding was necessary to avoid weldment cracking and residual austenite.

2.2 Attempts to raise steam conditions by use of austenitic steels

Low alloy and 12%CrMoVNb materials have provided a reliable basis for steam turbines operating with steam turbine inlet temperatures of 535-565°C, steam conditions which became standard during the period of the last 30-40 years. However they have insufficient creep strength for higher temperatures, at least unless used in very thick sections and with high levels of rotor cooling. In the early 1960s an attempt was made to introduce higher steam temperatures, as high as 649°C, through the exploitation of the greater creep strength of austenitic steels. However these steels present significant problems for application in thick section in boilers and steam turbines. Their high coefficients of thermal expansion, low thermal conductivities and low yield strengths result in poor resistance to thermal cycling. Instances of thermal fatigue, dimensional instability and structural collapse led in many cases to the downrating of those machines which were built. As a result of this poor experience, the attempt to raise standard operating temperatures above 565°C, at least in utility scale plant, was discontinued. Nonetheless several of these plants continued to operate. In the UK Drakelow C12, whose turbine was manufactured by GEC ALSTHOM, continued to operate with steam inlet temperatures of 593°C until very recently when it was finally retired after having achieved more than 120 000 hours in service.

3. THE DEVELOPMENT OF NEW 9-12%Cr STEELS

Through the 1960s and 1970s, although development focused on the optimisation and increased reliability of conventional materials, the possibility of improving turbine performance through exploitation of stronger high temperature materials was held under continuous review. The need was for materials with higher creep strength but physical properties and other mechanical properties at higher temperature which were equivalent to those of the conventional materials at the standard temperatures of 535-565°C. Towards the end of this period it was becoming clear that new alloys being developed in the USA and in Japan had potential for steam turbine applications.

3.1 Modified 9%Cr1Mo

In the mid 1970s ORNL were charged by the US Department of Energy with the development of a ferritic steel with high creep strength and good weldability for application in fast breeder reactors. They took as their basis the 9%Cr1%Mo alloy used for tubing in the UK and modified it through additions of V, Nb and N. The resultant alloy, known as modified 9%Cr1Mo, bore a superficial resemblance to the earlier generation of steels like FV448 but it differed in certain key respects. Nb content was reduced drastically, from the levels of 0.3% and above typical of FV448 to less than 0.1%. This eliminated the occurrence of large particles of primary NbC, not taken into solution during heat treatment, which contribute to the poor toughness of FV448. Carbon content was also reduced, to about 0.1%, contributing to improved weldability in two ways: by reducing hardness in the as-welded condition and by raising the martensite transformation temperatures so that the transformation of the weld metal to martensite is complete well above room temperature, avoiding the necessity for complex temperature control during welding typical of earlier alloys. The optimisation of alloy content also resulted in more stable long term creep properties. At 600°C the creep strength of the new alloy proved to be almost twice that of the previous generation of materials (6).

The growing awareness of the potential of this alloy during the 1980s coincided with intensified economic incentives for further reductions in construction and operation costs of power generation equipment. In addition a growing environmental awareness led to demand for a reduction in the environmental impact associated with power generation. A contribution to both the economic and environmental requirements can be made by increasing the efficiency of fuel conversion and it was clear that there was potential for achieving this by elevating steam temperatures and pressures through application of the new alloy. Furthermore the physical and mechanical properties of the new steel give higher resistance to thermal cycling than conventional low alloy materials so that no compromise in operational flexibility would be necessary in such high efficiency plant.

Work in the USA had focused on application of modified 9%Cr1Mo to pipework, associated fittings and tubing, and it has been standardised within ASME material standards and application codes where it is designated as grade 91. Additional development has been necessary to apply the material in the forms required for steam turbines. This company developed its application to castings through a number of programmes including an EPRI-sponsored programme (7) and through collaboration between the UK turbine makers and CEGB (8). GEC ALSTHOM worked with suppliers to carry out production trials, including the manufacture of a steam chest casting and thick section cast pipes, and the resultant castings were extensively characterised. The mechanical properties, including creep rupture and high strain fatigue properties, were determined and it was found that the creep rupture properties of the cast material closely matched those of the wrought material.

Development of fabrication procedures and characterisation of weldment properties has also been necessary for application of the steel to turbine components. Modified 9%Cr1Mo has been found to be easily weldable under a wide range of welding conditions including the highly restrained conditions representative of diaphragm welding. Weldments have been shown to have good impact strength and tensile strength and cross-weld creep rupture tests have shown the weldment strength to match that of the parent at temperatures approaching 600°C (Figure 3). In the long term, at this temperature and above, rupture strength is slightly reduced due to failure in a narrow soft zone at the outer edge of the weldment heat-affected zone, the type IV phenomenon common to all ferritic steels.

Initial work on fabrication of modified 9%Cr1Mo used consumables largely matching the chemical composition of the parent. However concern over hot cracking and over toughness properties led to the emergence of modified compositions. Under the consumable analysis specified by GEC ALSTHOM, Mn content is raised to avoid any risk of low melting point sulphur compounds leading to hot cracking in heavily constrained weldments. The highly effective dissolution of Nb achieved in the weld pool allows a slight reduction in Nb content in order to improve toughness without compromising creep strength. Small increases in Ni content also contribute to greater toughness. However the possibility of high Mn+Ni levels depressing the austenite transformation temperature to levels near the post weld heat treatment temperature (Figure 4) is also controlled.

Modified 9%Cr1Mo is now being applied to modern, high temperature machines. The 415MW turbines being supplied for the Skaerbaek and Nordjylland contracts in Denmark, which will operate with steam temperatures of 580°C, utilise this material for many cast components, including valve chests and inner-cylinders and also as forged plate and bar for welded diaphragms. The earlier work with suppliers ensured that no significant difficulties were encountered in the procurement of these materials (9).

A casting material similar to modified 9%Cr1Mo, but with slightly greater carbon content, has also been developed in Japan (10) and has been applied to turbines operating with steam temperatures of up to 593°C.

3.2 Rotor forgings

Modified 9%Cr1Mo was developed with fairly low yield strength. In most standards its yield or proof strength is specified as a minimum of 415MPa. While this is sufficient for tubing and pipework applications, higher strengths are required for rotor forgings.

Around the same time that modified 9%Cr1Mo was being developed in the USA, work began in Japan on materials suitable for rotor forgings. This work, which involved tests on a wide range of materials to find the best chemical analysis, led to the development of an alloy designated TMK1 (11). This steel is similar to modified 9%Cr1Mo but contains slightly more Cr (about 10%) and C (about 0.14%) and Mo is increased from 1.0 to 1.5%. The increase in C content, together with adoption of lower tempering temperatures, enables yield strengths greater than 700MPa to be achieved while at the same time fracture toughness is significantly better than in low alloy CrMoV materials. At 600°C the 100,000 hour creep strength of this alloy is similar to, or slightly greater than, that of modified 9%Cr1Mo.

GEC ALSTHOM took a leading role in initiating the corresponding European activities, a collaborative project between all the main European turbine makers under the auspices of COST 501. Since the beginning of this project, in the early 1980s, a very large number of small scale melts in different alloys and different heat treatment conditions have been assessed. Two of the most promising materials were applied to the manufacture of full scale rotor forgings which were sectioned for short and long term characterisation (12). These alloys were an optimised 10%CrMoVNbN alloy (Steel F) very similar to TMK1 and an alloy with an addition of 1% tungsten (Steel E).

The new steels are currently being applied to rotors in machines operating at high temperature and pressure. In Japan TMK1 has been applied at a number of power stations, as has an alloy containing additions of tungsten. In Europe, the COST developments have been used as the basis upon which GEC ALSTHOM selected 10%CrMoVNbN (Steel F) for application to the VHP and HP/IP rotors of the steam turbines for the Skaerbaek and Nordjylland power stations currently under construction in Denmark.

The development of modified 9%Cr1Mo, its application to a wide range of components including castings, and the development of similar materials for rotor forgings has now established a suite of materials suitable for steam turbines operating with steam inlet temperatures of up to at least 600°C (Figure 5). These materials are also suitable for applications at lower temperatures where their higher strength allows them to be used in thinner section than the conventional low alloy materials. They are thus economically competitive with the conventional steels and their improved high strain fatigue resistance coupled with their thinner section makes components in these materials more tolerant of thermal cycling. In the UK, modified 9%Cr1Mo has been used in steam plant operating at temperatures of 565°C to replace low alloy components susceptible to high strain fatigue cracking (13) and modified 9%Cr1Mo is becoming the standard material for main steam pipework even for machines operating at 540°C.

4. FURTHER DEVELOPMENT OF 9-12%Cr STEELS

With the development of the new 9-12%Cr materials described above the potential for improvements in the creep strength of this class of materials has been widely recognised. Further advances would enable the construction of still more efficient steam turbines so work is continuing on the development of these steels.

Within COST 501, GEC ALSTHOM is leading metallographic investigations to provide a fundamental understanding of the creep resistance of these steels [14] so that further advances in creep strength can be made. Already many largely empirical attempts have been made, or are in progress, to further improve the creep strength of these steels. These attempts are principally centred around two principles, the addition of tungsten in substitution of molybdenum and the addition of boron.

4.1 The use of tungsten to elevate creep strength

The intellectual basis for this approach was laid by Fujita. He reported that creep strength was optimised when the Mo+0.5W content was around 1.5%. TMK1 obeys this principle with a Mo content of 1.5% and no tungsten. However he also reported improved creep strength when W rather than Mo was used (15).

Some of the rotor materials described above already contain up to 1%W but such additions have not resulted in creep strength discernibly greater than that of the W-free compositions. Similarly, in an attempt to raise creep strength of cast materials, an alloy containing 1% tungsten was successfully applied to the manufacture of a valve chest casting within the COST 501 project (16). Long term characterisation of this casting is currently in progress. Although the chemistry and heat treatment of the alloy give it enhanced tensile strength in comparison with modified 9%Cr1Mo and hence an advantage in short-term creep strength (17), its advantage in the long term has still to be established.

More generous additions of tungsten have sometimes been more successful. One of the most mature developments is that of NF616, containing about 1.8%W and 0.5%Mo. This alloy has 100,000 hour strength at 600°C estimated to be about 30% greater than that of the tungsten free alloys and has already gained acceptance under ASME codes for pipework applications (18). On the other hand an alloy with similar W level investigated within COST 501 showed poor long term strength. Although a rotor forging with 1.8%W has been produced in Japan (19) it seems to offer at best only a very marginal advantage in creep strength over the W free alloys like COST 501 Steel F and TMK1.

The introduction of large amounts of W encourages the formation of delta ferrite. This is avoided in NF616 by the limitation of Cr content to around 9%. However if alloys like these are to be applied at temperatures above 600°C then oxidation resistance becomes a concern and it becomes desirable to increase Cr content to nearer 12%. The combination of such Cr levels with high W content would certainly result in significant quantities of delta ferrite so elements which increase the austenite stability field and hence suppress delta ferrite, such as cobalt or copper, are added.

In Japan a 12%Cr alloy containing 2.5%W balanced by 2.5%Co, designated HR1200, has been developed for rotor forgings (20). Essentially the same alloy is also being developed for tubing applications under the designation NF12 (21).

The mechanism by which high W content enhances creep strength has not been established and, despite data of over 50,000 hours in duration on NF616, doubts remain over its long term performance. In steels containing Mo of around 1% or above significant quantities of Laves phase, $Fe_2(Mo,W)$, are formed on long term exposure at the potential service temperatures around 600°C. However the scale of this precipitation is intensified when tungsten is added to the steel. In COST 501 detailed quantitative metallography has been carried out on creep specimens from a range of alloys. Measurements of Mo and W in solid solution in steels containing levels of these elements leading to similar values of (Mo+0.5W) but with different W contents (22) are shown in Figure 6. In the alloy with 1.8%W the matrix is denuded of its alloy content to a much greater extent than in the case of the W free alloy, reflecting the more intense precipitation of Laves phase. Hald has suggested that this precipitation process is responsible for enhanced creep strength. However he has also established the kinetics of this precipitation and shown that the precipitation process is complete after about 30,000 hours at 600°C (23). After that time the absence of further precipitation and the rapid coarsening rate of Laves phase gives concern over the longer term creep strength.

4.2 The use of boron to elevate creep strength

Work on steels containing B dates back to the origins of the COST 501 project in the early 1980s. It soon became apparent that large additions of this element led to severe segregation and an optimum level of around 100ppm was established. A steel containing this level of B was further developed in later rounds of COST 501 and showed promise of creep strength slightly greater than that of the B free steels (12). A VGB funded project is about to apply this steel to the manufacture of a full scale rotor forging.

The use of boron has also been adopted in Japan. In addition to 2.5%W, the HR1200 alloy contains significant levels of boron. In line with the European experience, early work on this alloy with levels of B around 180ppm resulted in segregation and later development has focused on reduced B levels, around 100ppm (20). NF616 also contains some B, typically about 30ppm, and it is uncertain what contribution this element, rather than the high W content, makes to its high creep strength.

As in the case of W the mechanism by which B enhances creep strength is not established. Detailed metallographic studies have shown this element to concentrate in $M_{23}C_6$ carbides (24) and it is possible that it modifies the coarsening behaviour of these particles. From what little is known there is no reason to fear that such a mechanism would lead to some long term instability as is feared for tungsten.

In order to resolve these uncertainties and in an attempt to establish even stronger alloys with good oxidation resistance, development is continuing around the world on alloys with 12%Cr and high levels of W balanced by elements like Co or Cu and with additions of B. In COST 501 a series of trial melts is being tested and metallographic investigations are continuing to throw light on the mechanisms of creep resistance. The success of these developments would enable the design and construction of steam turbines with inlet temperatures approaching 650°C.

A third strategy for improvement of creep strength in 9-12%Cr steels is centred around the generation of a very stable dispersion of TiN or TiC particles. However because of the thermodynamic characteristics of such particles it is necessary to employ sophisticated processing routes to develop the required particle dispersions, involving either reactive powder metallurgy (25) or carefully controlled hot working techniques (26). These techniques are in the early days of development and may not be suitable for many of the larger steam turbine components. Nonetheless the long term potential of such routes has yet to be fully investigated.

5. APPLICATION OF NICKEL-BASED ALLOYS

Even the most optimistic view does not suggest that the 12%Cr alloys will enable steam turbines operating at temperatures greater than 650°C. Further increases in operating temperature, and hence efficiency, will require the application of a new family of steam turbine materials.

In Japan a rotor in austenitic stainless steel, modified A286, has been manufactured and has seen exposure in a test turbine (27). However it is likely that the austenitic stainless steels would offer only a marginal advantage in creep strength over the most advanced 12%Cr steels. Given the difficulties associated with the poor thermal cycling capabilities of the austenitic steels, coupled with their increased cost, it is far from clear that the marginal increase in efficiency which might be possible through application of these steels will be sufficient to justify their use.

A more attractive alternative is the use of Ni-based superalloys for the highest temperature components of both boilers and turbines. The mechanical and physical properties of these alloys make them much more tolerant of thermal cycling than are the austenitic steels and the creep strength of even the most easily manufactured and fabricated but weakest of these alloys is sufficient to allow operation with steam temperatures of 700°C or more (Figure 7). Under these conditions, the conventional combination of pulverised coal-fired boilers and steam turbines in the Rankine cycle has potential for efficiencies greater than 55%, a level which compares favourably with the projected efficiencies for more novel coal-burning technologies under consideration, such as coal gasification or pressurised fluidised bed concepts. Although the costs of these materials are significantly higher than those of materials currently used, the restriction of their application to critical high temperature components will mean that the impact of this increased cost on the plant as a whole will not be prohibitive and will be more than compensated by the lower operating costs following from the increased efficiencies enabled by application of these materials.

Currently the major applications of Ni-based alloys are in aerospace and gas turbine technologies. For steam turbines the largest forgings and castings which can be produced will be required but these will be smaller than the forgings and castings available in conventional materials. Therefore novel steam turbine architectures and construction technologies will be exploited. A project to develop and demonstrate such plant, involving all major European power generation plant manufacturers, material suppliers and utilities, and to be supported by the European Commission is currently being initiated. GEC ALSTHOM is taking a leading role in the activities of the turbine makers. The materials development necessary to support this project will include demonstration of full scale components, development of joining technology and generation of long term data. In order to have this technology in place when required for plant design and construction, materials development will be undertaken from the beginning of the project.

6. CONCLUSION

Metallurgical development around the world, in which GEC ALSTHOM has played a significant and sometimes leading role, has led to the availability to the turbine engineer of a range of 9-12%Cr materials suitable for the full range of steam turbine components, enabling operation at temperatures of up to around 600°C and significant gains in efficiency without any need to compromise on operational flexibility. These materials are already being used for construction of new plant, both in Europe and in Japan. Further development is required, and is in progress, on more advanced 9-12%Cr alloys to allow operation at temperatures up to 650°C.

For even higher temperatures, of up to 700°C and beyond, it will be necessary to apply nickel based materials and the foundations for the necessary development of these materials are currently being laid. The gains in efficiency which will be possible through exploitation of these materials will bring about a reduction in fuel consumption of more than 30% with all the associated economic and environmental benefits. However limitations on forging and casting size will require the application of radical turbine designs and the turbine engineer cannot rely on material developments alone if such benefits are to be realised.

References

1. Greenfield, P., Thornton, D.V. 'Integrity of steam turbine rotor, chest and casing materials', Proc. of I Mech E Conf on Steam Turbines for the 1980s, London, 1979.
2. Thornton, D.V. 'Materials for modern high temperature steam turbines', Proc. of I Mech E Conf on Steam Turbines for the 1990s, London, 1990.
3. Norton, J.F., Strang, A. 'Improvement of creep and rupture properties of large 1%CrMoV steam turbine rotor forgings', JISI, Feb 1969.
4. Price, A.T. 'Rupture ductility of creep resistant steels fabrication and welding' Proc of Institute of Metals Conf on Rupture Ductility of Creep Resistant Steels, York, December 1990.
5. Wickens, A., Strang, A., Oakes, G. 'High temperature properties of creep resistant 12Cr steam turbine blading steels', Journal of the I Mech E, (1980), 11-18.
6. Sikka, V.K., Cowgill, M.G., Roberts, B.W. 'Creep properties of modified 9Cr-1Mo steel', Proc. Conf. 'Ferritic alloys for use in nuclear energy technologies', Snowbird, Utah, June 1983.
7. Mayer, K.H., Gysel, W., 'Modified 9%CrMo cast steel for casings in improved coal-fired power plants', Proc. of 3rd EPRI Int. Conf. on Improved Coal-Fired Power Plants (ICPP), San Francisco, April 1991.
8. Thornton, D.V., Hill, R. 'The fabrication and properties of high temperature, high strength steel castings', ibid.
9. Thornton, D.V., Taylor, M. 'Experience in the manufacture of steam turbine components in advanced 9-12%Cr steels', this conference.

10. Yamada, M., Watanabe, O., Tsunoda, E., Miyazaki, M. '12Cr heat resistant steel castings for advanced steam turbines', Proc. of 2nd EPRI Int. Conf. on Improved Coal-Fired Power Plants (ICPP), Palo Alto, November 1988.
11. Nakabayashi, Y. et al 'Advanced 12Cr steel rotor (TMK1) for EPDC's 50MW high temperature turbine step 1 (593°C/593°C)', Proc. of 1st EPRI Int. Conf. on Improved Coal-Fired Power Plants (ICPP), Palo Alto, November, 1986.
12. Berger, C., Scarlin, R.B., Mayer, K.H., Thornton, D.V., Beech, S.M. 'Steam turbine materials: high temperature forgings', Proc. of COST 501 Conf. Materials for Advanced Power Engineering 1994, Liège, October 1994.
13. Greenwell, B.S., Taylor, J.W. 'The properties of candidate welds in 9CrMoNbV steel', Proc. I Mech E Conf. Steam Turbines for the 1990s, London, 1990.
14. Vanstone, R.W. 'Microstructure and creep mechanisms in advanced 9-12%Cr creep resisting steels. A collaborative investigation in COST 501/3 WP11', Proc. of COST 501 Conf. Materials for Advanced Power Engineering 1994, Liège, October 1994.
15. Fujita, T., Asakura, K., Sawada, T. 'Creep rupture strength of low C, 10Cr-2Mo heat resisting steels', Metallurgical Transactions, 1981, 2A, 1071.
16. Scarlin, R.B., Berger, C., Mayer, K.H., Thornton, D.V., Beech, S.M. 'Steam turbine materials: high temperature castings', Proc. of COST 501 Conf. Materials for Advanced Power Engineering 1994, Liège, October 1994.
17. Thornton, D.V., Vanstone, R.W. 'Materials developments for application in steam turbines for fossil fired plant', 3rd Int Charles Parsons Conference 'Materials engineering in turbines and compressors', Newcastle upon Tyne, April 1995.
18. EPRI/National Power conference, 'New steels for advanced plant up to 620°C', London, May 1995.
19. Kamada, M. et al 'An advanced turbine rotor of 12Cr steel (TMK2) applicable to 600°C steam temperature', 3rd Int Charles Parsons Conference 'Materials engineering in turbines and compressors', Newcastle upon Tyne, April 1995.
20. Hidaka, K. et al, 'Development of large scale 12Cr ferritic steel turbine rotor aiming the application of ultra supercritical steam power plant at 650°C', ibid.
21. Fujita, T. 'Future ferritic steels for high temperature service', EPRI/National Power conference, 'New steels for advanced plant up to 620°C', London, May 1995.
22. Foldyna, V. COST 501/3 WP11 Project 11CS5.
23. Hald, J. 'Materials comparison between NF616, HCM12A and TB12M - microstructural stability and ageing', EPRI/National Power conference, 'New steels for advanced plant up to 620°C', London, May 1995.
24. Lundin, L., Norell, M., Andren, H.O., Nyborg, L. 'Remanent life assessment of creep resistant modified 12% chromium steels: microstructural analysis and microstructural development models' (to be published).
25. Hamerton, R.G., Jaeger, D.M., Jones, A.R. 'Titanium nitride and nitrogen strengthened stainless steels', Proc. of COST 501 Conf. Materials for Advanced Power Engineering 1994, Liège, October 1994.
26. Buck, R.F., Garrison, W.M. 'Creep resistant martensitic steel', Advanced Materials and Processes, August 1996.
27. Fujita, A. et al 'Application of modified A286 iron base superalloy to USC turbine rotor', Proc. of COST 501 Conf. Materials for Advanced Power Engineering 1994, Liège, October 1994.

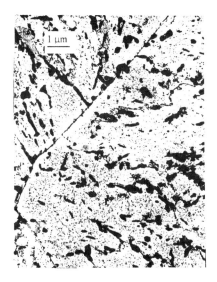

Fig. 1. Optimised microstructure in 1%CrMoV rotor forgings

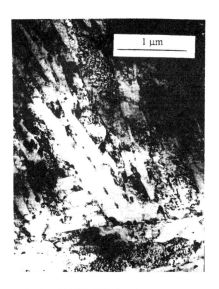

Fig. 2. 12%CrMoVNb microstructure

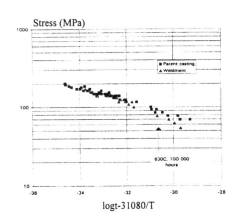

Fig. 3. Rupture strength of modified 9%Cr1Mo castings and weldments

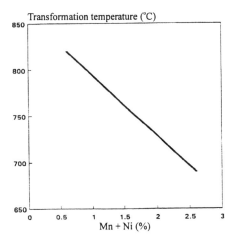

Fig. 4. Influence of Mn and Ni in modified 9%Cr1Mo weld metal on austenite transformation temperature

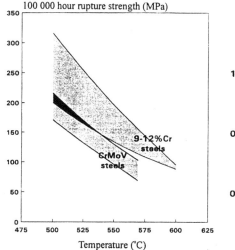
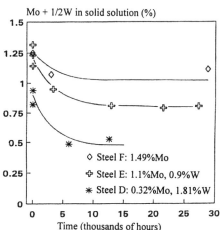

Fig. 5. Improved rupture strength of 9-12%Cr steels now available

Fig. 6. Reduction in Mo and W in solid solution due to formation of Laves phase during creep at 600°C

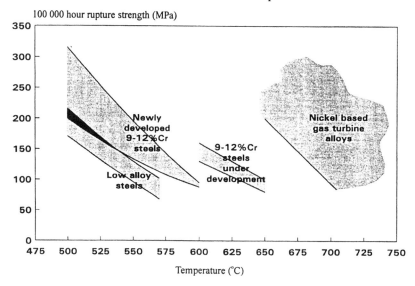

Fig. 7. Rupture strength of nickel based alloys and expected strength of 9-12%Cr steels under development compared with low alloy steels and recently developed 9-12%Cr steels

C522/008/97

Development of 9–12% Cr steels for all-ferrite steam turbine at target temperature of 650°C

K HIDAKA PhD, JInst Met, **Y FUKUI** ASM, TMS, and **R KANEKO** BSc, MJSME
Hitachi Limited, Japan
T FUJITA
University of Tokyo, Japan

Introduction

New 12% CrWCoB steel is studied for the application of the highly efficient ultra supercritical pressure steam (USC) power plant because of the protection of natural environment and the preservation of fossil fuel. The conventional ferritic steels are no longer applicable for the USC rotor because of their low level of creep rupture strength at operating temperature above 600°C. The austenitic steel (A286) became the rotor material's candidate[1], but difficulty arises for the enlargement of rotor scale due to strong resistance against upsetting and the large thermal expansion coefficient. Application of ferritic steel is revived for USC power plant by new addition of boron (B) and cobalt (Co) to the 12% chromium (Cr) steels[2-4]. Our previous research completed the development of 12% CrWN steel rotor[5] named HR1100 (Hitachi Developed Rotor Forging for 1100°F). The installation of HR1100 is already scheduled for 1000 MW class commercial USC power plant.

The development of 12% CrWCoB steel is aiming the application of 650°C class USC power plant whose rotor named HR1200 (Hitachi Developed Rotor Forging for 1200°F) and blade named TAF650 (The University of Tokyo, Akutagawa and Fujita for 650°C) in future. The casing material is also studied the application of 9% CrWN steel[6] for the future USC power plant.

(1) Determination of chemical composition of the 12% CrWCoB steel

New alloying elements, B and Co, are chosen for 12% CrW ferritic steel which is suggested by Fujita[7] because addition of B increases hardenability and that of Co suppresses delta ferrite formation. The various compositions of alloy steel are prepared by vacuum casting ranging from 10 to 150 kg ingots to find the optimum amount of alloying elements in terms of mechanical properties. It is said that the addition of the ferrite former elements, e.g. Mo, W, vanadium (V), and niobium (Nb) to 12% Cr steel is necessary to increase the creep rupture strength. However, excess amounts of ferritic former addition may cause delta ferrite formation because the ferrite forming elements stabilize ferrite rather than austenite in phase diagram. Therefore, addition of austenite formers is required to suppress the stability of ferrite. However, the most austenite formers decrease the creep rupture strength. Co is a good candidate among the austenite formers because addition of Co increases the creep rupture strength of the 12% Cr steel[8]. As the 12% CrWCoB steel contains large amount of the ferrite formers, addition of Co is necessary to prevent delta ferrite formation.

The typical alloying elements, carbon (C) and nitrogen (N), are also studied to find the optimum amount of addition for the 12% CrWCoB steel. Table 1 shows the chemical composition of examined ingots prepared by vacuum induction melting. Effect of nitrogen is studied by the ingots from number 1 through 3. The sample number 4 through 6 is used for the study of Co on the creep rupture strength. Effect of B is also studied in terms of the creep rupture strength from ingot number 7 to 9.

Effect of C

Carbon is essential element to control the mechanical properties of the steels. It is said that precipitation hardening of steel is based on the formation of fine carbides in addition to the martensite formation caused by dissolved C. The size of precipitates is important for the mechanical properties because the fine precipitates introduce precipitation hardening and coarse precipitates cause brittle fracture. However, carbon addition may promote the deterioration of microstructure during creep examination because fast coarsening of precipitates takes place at elevated temperatures during creep examination. Fig. 1 shows the negative slope of creep rupture strength as a function of C content. Low carbon content is recommended for the better creep rupture strength of the 12% CrWCoB steel.

Effect of N

Nitrogen addition has been positively promoted up to the solubility limit of the 12% CrW steel. However, simultaneous addition of N and B may form the boron nitride which degrades the mechanical properties, especially in ductility. Fig. 2 shows negative slope of N addition to creep rupture strength. N content to the 12% CrWCoB steel is allowed only from the atmosphere during manufacturing process. The normal level of nitrogen is about 0.02wt% in the 12% CrWCoB steel from the atmosphere.

Effect of B

B is not easily detected by conventional analyzer because the atomic number and the amount of addition are small. Fig. 3 shows the positive slope of B addition for creep rupture strength. The creep rupture strength is significantly increased by small amount of B addition in B free steel. The slope of creep rupture strength becomes gradual increases at relatively large amount of B addition.

Effect of Co

Although the creep rupture strength is increased with the amount of Mo, W, V, and Nb increases, large amount of ferrite former addition causes delta ferrite formation which may cause the brittle fracture. The delta ferrite formation is completely prohibited for the components of rotating structure. Thus, austenite former addition is required to prevent delta ferrite formation. The positive and linear response of creep rupture strength is obtained as a function of Co content as shown in Fig. 4.

When chemical composition of the structural components for USC power plant is determined, large-scaling of structural material is the next step to evaluate applicability of the 12% CrWCoB steel for the target temperature 650℃.

(2) Trial manufacturing of 10 tons class ingot forging

Trial manufacturing of 10 tons ingot forging is carried out by the results obtained from the optimization of alloy composition. The ingot chemistry is shown in Table 2. The ingot is prepared by electro-slag remelting (ESR) after ladle refining. Segregation of alloy elements is not found in whole area of the ingot forging.

The heat treatment is studied to find optimum mechanical properties of the 12% CrWCoB steel as shown in Fig. 5. The mechanical strength is decreased as the second tempering temperature increased because of the softening effect of the second tempering. However, fracture appearance temperature transition (FATT) is improved as the second tempering temperature increased by the same reason of softening. The optimun heat treatment for this material is the quenching at 1050℃ and the second tempering at 700℃.

(3) Trial manufacturing of 20 tons class ingot rotor shaft

The same process as the 10 tons class ingot manufacturing is carried out for the trial manufacturing of 20 tons class rotor whose weight is similar to high pressure rotor shaft for 1000 MW class steam turbine power plant. Binary slag, CaF_2-Al_2O_3, was used for the ESR process.

The chemical composition of 20 tons rotor is listed in Table 3 with the comparision of the 12% CrWN steel. Forging temperature is carefully chosen by the results obtained from the reduction of area of the specimens examined at elevated temperatures because grain boundary may be reduced to a liquid state due to the eutectic reaction of Fe_2B and $Fe^{(1)}$. The forging temperature is limited up to 1130°C because reduction of area becomes less than 50% above 1130°C as shown in Fig. 6.

(4) Determination of heat treatment for 20 tons class ingot rotor shaft

The specimens were machined from the corner of the drum girth for the determination of the heat treatments of the rotor. Fig. 7 shows the creep rupture strength of quenching effect as a function of quenching temperature. The creep rupture strength shows positive slope against quenching temperature. It can be concluded that higher quenching temperature causes better creep rupture strength. However, the grain coarsening due to the exposure at elevated temperatures during quenching process may cause the degradation of toughness. Fig. 8 shows the effect of second tempering temperature on creep rupture strength. The creep rupture strength shows negative slope against second tempering temperature. The heat treatment is determined by the balance of strength and ductility introduced from quenching and tempering, respectively.

The 20 tons rotor shaft is quenched from 1050°C and the secondary tempered at 700°C.

(5) Creep rupture strength of the rotor

Fig. 9 shows the rotor shaft which is machined after heat treatment. The specimens are prepared from centre core in tangential direction and drum girth in radial direction after heat treatment. The evaluation of creep rupture strength of the rotor shaft is shown in Fig. 10. The constant of Larson-Miller Parameter (LMP) plots is given by C=40 for the 12% CrWCoB steel rotor. The results of the creep rupture strength at 10^5 hours are estimated from the acceleration tests at higher temperatures. The creep rupture strength from the LMP plots satisfies the target value, above 100 MPa at 650°C for 10^5 hours. The creep rupture strength of both girth and core shows similar value, which implies the homogeneity of the structure from core to girth because of the utilization of the ESR technique. The microstructure of the rotor is confirmed as 100 % tempered martensite. The advantage of creep rupture strength of the 12% CrWCoB steel is significantly higher than that of the 12% CrWN steel as shown in Fig. 11 with comparision of austenitic alloy, A286.

The 12% CrWN steel, HR1100, is already scheduled for the application of 600°C class USC power plant. The creep rupture strength of HR1100 shows 120 MPa at 600°C from the LMP plots. The creep rupture strength of HR1200 shows the same stress level at 650°C. It can be concluded that the addition of Co and B to 12% Cr steel may realize the application of 12% Cr ferritic steel to the USC power plant operating at 630~650°C in future.

(6) Manufacturing of blade material

Blade material of 12% CrWCoB steel is forged after ingot preparation by vacuum induction melting. Creep rupture strength of the blade material is evaluated by the LMP plots. Chemical composition of the blade material is listed in Table 4. The creep rupture strength shows satisfactory result, above 100 MPa at 650°C for 10^5 hours, as shown in Fig. 12.

(7) Determination of chemical composition of cast steel

Chemical composition of the cast material is modified from 9% CrWN steel (NF616) as listed in Table 5. The modification is focused on Ni content because Ni addition improves mechanical properties such as tensile strength and ductility at room temperature. Ni content is increased up to 1 wt%. Fig. 13 shows positive responce of the mechanical strength as a function of Ni content. Ni content is determined as 1 wt% for the casing material.

(8) Trial manufacturing of casing for high pressure rotor

Trial manufacturing of cast casing for high pressure turbine is carried out to investigate the castability and mechanical properties. Chemical composition of the high pressure casing is listed in Table 6. The Ni content was lower than the target value.

Fig. 14 shows the cutaway of the high pressure turbine inner casing. One-quarter size of the high pressure turbine casing is prepared. No significant segregation is observed by macroscopic examination. Charpy impact, high temperature tensile and creep rupture strength tests are studied. Impact absorbed energy and FATT are listed in Table 7. Charpy impact test shows satisfactory results. The results of the high temperature tensile test and the creep rupture strength are shown in Fig. 15 and 16, respectively. The results shows satisfactory properties for inner casing of high and medium preesure turbine.

Conclusions

The following conclusions are obtained from the experimental results.
(1) The addition of Co and B to 12% CrW steel significantly increases the creep rupture strength.
(2) The creep rupture strength of 20 tons class rotor and blade material from the LMP plots satisfies the target value, 100 MPa at 650°C for 10^5 hours.
(3) No significant segregation of alloying elements is detected in HR1200 rotor forging manufactured by the ESR technique.
(4) 9% CrWN steel is applicable to casing for future USC power plant.

Acknowledgements

One of the authers ackowledges Dr. M. Shiga and Mr. N. Morisada for useful discussion about the trial manufacturing of the 20 tons class rotor forging.

References
(1) C. W. Elaton and R. Sheppard : Trans. ASME, Vol. 79, No. 2, (19957) 417-426
(2) K. Hidaka et al.: Materials for Advanced Power Engineering, Part I, 281-290, Liège, Belgium 1994, Kluwer Academic Publishers.
(3) K. Hidaka et al.: Materials Engineering in Turbines and Compressors, Part I, 191-200, Newcastle, UK 1995, The Institute of Materials.
(4) T. Uehara et al. : Materials Engineering in Turbines and Compressors, Part II, 391-400, Newcastle, UK 1995, The Institute of Materials.
(5) F. Ito et al. : Inter. Conf. ASME, 86-JPGC-Pwr-3, Portland, Oregon, U.S.A., Oct. 19-23, 1986.

(6) H. Miura et al. : Materials for Advanced Power Engineering, Part I, 361-372, Liège, Belgium 1994, Kluwer Academic Publishers.
(7) T. Fujita: Materials Engineering in Turbines and Compressors, Part II, 493-516, Newcastle, UK 1995, The Institute of Materials.
(8) F. B. Pickering : Physical Metallurgy and the design of steels, Applied Science Publishers Ltd., 1978.

Table 1 Chemical composition and thermal history of sample ingots

Sample No.	Chemical composition (wt%)							Thermal history	Studied Element
	C	Cr	Mo	W	N	Co	B	Cr eq.		
1	0.12	11.05	0.15	2.64	0.047	2.94	0.014	3.2	1050°C×5h 100°C/h	
2	0.11	10.98	0.15	2.59	0.025	2.87	0.014	4.4	570°C×20h A.C.	Nitrogen
3	0.11	10.98	0.15	2.62	0.025	2.93	0.014	4.2	720°C×20h F.C.	
4	0.11	11.11	0.18	2.51	0.017	---	0.014	10.2	1050°C×5h 100°C/h	
5	0.11	11.02	0.17	2.54	0.018	1.46	0.013	7.2	570°C×20h A.C.	Cobalt
6	0.11	11.05	0.17	2.53	0.019	2.92	0.014	4.3	690°C×20h F.C.	
7	0.10	10.80	0.19	2.54	0.025	2.51	---	5.0	1050°C×5h 100°C/h	
8	0.10	10.94	0.20	2.52	0.022	2.51	0.008	5.3	570°C×20h A.C.	Boron
9	0.10	10.88	0.20	2.52	0.023	2.51	0.017	5.3	690°C×20h F.C.	

Table 2 Chemical composition of 10 tons class ingot forging

C	Si	Mn	Ni	Cr	Mo	V	W	Nb	Co	B	N
0.12	0.03	0.48	0.52	11.03	0.21	0.22	2.68	0.08	2.59	0.007	0.021

(wt%)

Table 3 Chemical composition of 20 tons class ingot forging and HR1100 rotor

	C	Si	Mn	Ni	Cr	Mo	V	W	Nb	Co	B	N
HR1200	0.09	0.02	0.53	0.51	11.00	0.23	0.22	2.66	0.07	2.53	0.018	0.020
HR1100	0.12	0.06	0.52	0.60	10.22	1.17	0.17	0.38	0.05	—	—	0.050

(wt%)

Table 4 Chemical composition of blade material (TAF650)

C	Si	Mn	Ni	Cr	Mo	V	W	Nb	Co	B	N
0.11	0.01	0.47	0.47	10.98	0.15	0.19	2.62	0.08	2.91	0.015	0.020

(wt%)

Table 5 Chemical composition of 50 kg ingot

C	Si	Mn	Ni	Cr	Mo	V	W	Nb	B	N	O
0.13	0.22	0.49	0.57	9.59	0.58	0.16	1.71	0.07	0.0021	0.0434	0.0037
0.12	0.22	0.49	1.09	9.66	0.58	0.17	1.72	0.07	0.0018	0.0446	0.0049

(wt%)

Table 6 Chemical composition of casing for high pressure turbine

C	Si	Mn	Ni	Cr	Mo	V	W	Nb	B	N	O
0.13	0.17	0.47	0.60	9.55	0.52	0.22	1.66	0.06	0.0016	0.0541	0.0040

(wt%)

Table 7 Charpy impact test results

Absorbed enrgy (kgf-m)		FATT (°C)
vE0	vE20	
1.5	2.1	76
1.6	2.1	

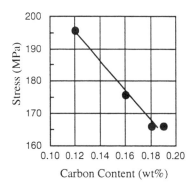

Fig. 1 Creep rupture strength of 10^4 hours from LMP plot as a function of C content.

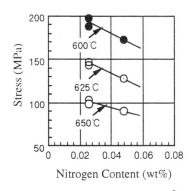

Fig. 2 Creep rupture strength of 10^5 hours from LMP plot as a function of N content.

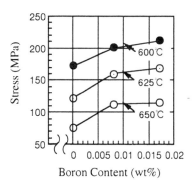

Fig. 3 Creep rupture strength of 10^5 hours from LMP plot as a function of B content.

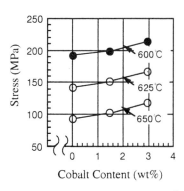

Fig. 4 Creep rupture strength of 10^5 hours from LMP plot as a function of Co content.

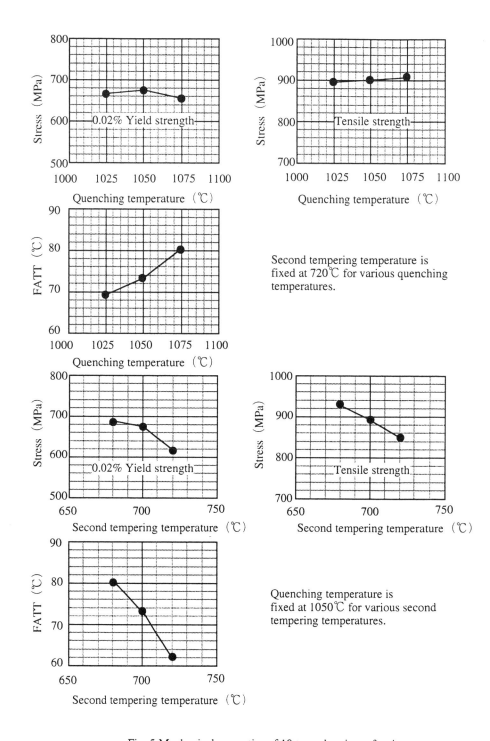

Fig. 5 Mechanical properties of 10 tons class ingot forging as a function of heat treatments

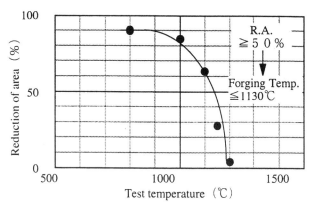

Fig. 6 Evaluation of forgeability by high temperature tensile tests

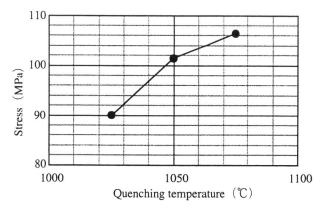

Fig. 7 Determination of quenching temperature in terms of creep rupture strength at 650°C for 10^5 hs as a function of quenching temperature

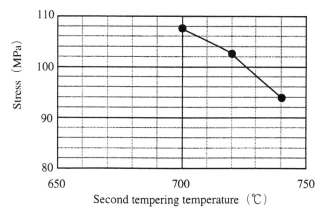

Fig. 8 Determination of second tempering temperature in terms of creep rupture strength at 650°C for 10^5 hs as a function of second tempering temperature

Fig. 9 Heat-treated 20 tons class rotor shaft after machining

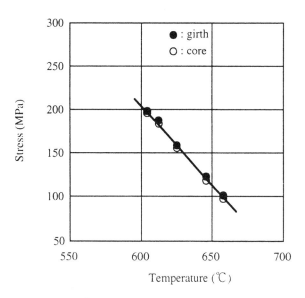

Fig. 10 10^5 hs creep rupture stress of 20 tons class rotor forging evaluated by LMP plot

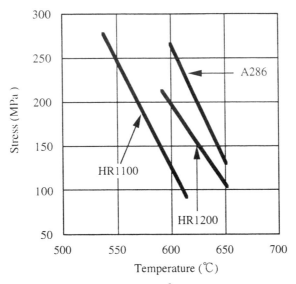

Fig. 11 Comparision of 10^5 hs creep rupture strength of rotor forgings

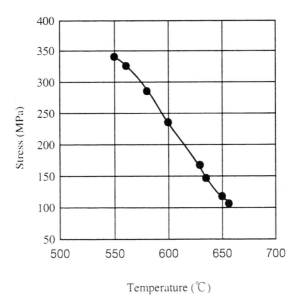

Fig. 12 10^5 hs creep rupture strength of blade material (TAF650) evaluated by LMP plot

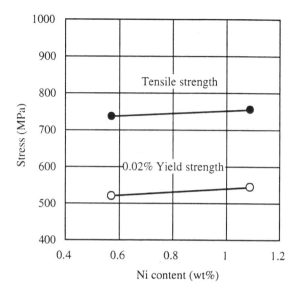

Fig. 13 Mechanical strength as a function of Ni content for 9% CrWN steel casting

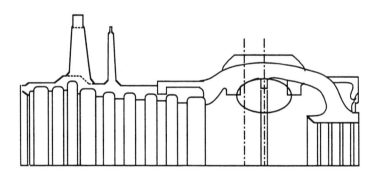

Fig. 14 Cutaway of the high pressure turbine casing

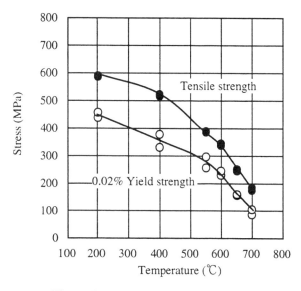

Fig. 15 Tensile examination of 9% CrWN steel casing at elevated temperatures

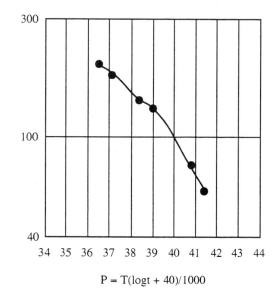

$P = T(\log t + 40)/1000$

Fig. 16 Creep rupture strength of 9% CrWN steel casing in terms of Larson-Miller Plot

C522/012/97

Development of 11CrMoWCo heat resistant steel for fossil thermal plants

M OHGAMI, Y HASEGAWA, and H NAOI
Steel Research Laboratories, Nippon Steel Corporation, Chiba, Japan
T FUJITA
Faculty of Technology, University of Tokyo, Japan

Synopsis We successfully developed a 11%Cr ferritic heat resistant steel, named NF12, which has good high temperature creep strength compared with conventional heat resistant steels. The representative chemical composition of NF12 consists of 11% chromium, 2.6% tungsten, and about 2% cobalt. Its extrapolated creep rupture strength reaches about from 1.6 to 1.9 times as high as that of Modified 9Cr-1Mo steel. The estimated creep properties depends on the estimation methods. We hope the remarkable improvement of steam condition for the Ultra Super Critical power plant being under consideration through the development of NF12 for thin wall components application.

1. Introduction

There is a growing tendency for the fossil fired power plants to increase the steam condition in order to improve the heat exchanging efficiency from the view point of energy saving in recent years. This trend is also a desirable direction because the amount of CO_2 emission from power generation plant must be decreased at the same time for the environmental protection. Such increase of steam condition is only achieved by the development of new martensitic heat resistant steel, which equips the far better mechanical properties than those of conventional ferritic steels at elevated temperature.

As a result of author's research activities on the tungsten containing 9% chromium martensitic steel, we developed the heat resistant steel[1]~[4], NF616, with the superior creep properties to those of the Mod.9Cr-1Mo steel (ASME T91/P91). This creep resistant steel indicates about 1.3 times as high creep rupture strength as that of ASME T91/P91 steel. But this creep property does not satisfy the latest designed steam condition, up to 650°C and 350 bars. Therefore, the new martensitic steel with higher creep properties was craved and have been researched for this purpose.

Fig.1 is a ferritic materials application map in terms of steam conditions for fossil power plant[5]. 2.25Cr-1Mo and X20CrMoV121 steels are applicable at critical or at super critical steam condition, but the application of those materials at Ultra Super Critical steam condition is impossible. Even in case of ASME T91/P91 steel, the creep properties of the steel are not enough to apply when the steam temperature reaches about 600°C. NF616 is available even if the steam condition is much more severe, and up to 620°C. NF12 is the candidate material for the boiler application at higher steam temperature.

2. Alloy design

The alloy design concept for new 11% chromium containing steel realizes the higher creep properties and mechanical properties than those of NF616 at elevated temperature. The chemical composition of the 11% chromium steel, so-called NF12, was determined as follows;

C and N : The better balance of carbon and nitrogen contents, 0.08% and 0.05% respectively, than that of NF616, 0.1% carbon and 0.05% nitrogen, increase the creep rupture strength for NF12.

Cr : Chromium content of NF12 is 2% higher than that of NF616 in order to obtain the enough oxidation resistance and corrosion resistance at about 650°C.

Ni : According to the investigations of effect of nickel content to the creep rupture strength, higher nickel than 0.5% content in ferritic steel decrease the creep rupture strength remarkably after long time exposure at from 550°C to 650°C. But small amount of nickel addition, it must be lower than 0.5% and as low as possible, decrease the retainment of delta ferrite phase and mitigates the deterioration of toughness owing to the increase of chromium content.

W and Mo : Based on the survey of tungsten and molybdenum compositions balance, the 1.5 for Molybdenum Equivalent with higher tungsten content, which equivalent is defined as the value of molybdenum content plus half of tungsten content, obtains the highest creep rupture strength for 9 to 12% chromium steel at 650°C. Therefore NF12 contains 2.6% tungsten and 0.15% molybdenum in it.

Nb and V : Nb and V increase the creep properties through the precipitation strengthening effect by the carbides and nitrides precipitation. 0.07% niobium and 0.2% vanadium is the best compositions for the creep rupture strength.

Co : Cobalt is well-known austenite phase former in chromium containing ferritic steel. 2.5% cobalt completes the full-martensitic micro-structure and increase the hardenability. The higher than 3% cobalt content decrease the creep rupture strength remarkably at 650°C for long time creep exposure.

B : From 0.003 to 0.004% boron strengthen the primary austenitic grain boundary and increase the creep rupture strength at 650°C. The too much boron, which is higher than 0.01%, deteriorates the weldability of the steel.

Si : Decrease of silicon content increase the toughness and creep rupture elongation. But 0.2% silicon is necessary to the enough steam oxidation resistance at about 650°C.

Mn : Manganese is also the effective austenite former and therefore completes the full martensitic microstructure and increases the hardenability. It affects the creep properties almost like nickel, therefore it is restricted 0.5% at most.

Typical chemical composition of NF12 is listed in Table 1 comparing with those of conventional ferritic steels.

3. Chemical compositions and their manufacturing processes.

The chemical compositions of specimens for the experiments are shown in Table 2. The parent alloys were melted in vacuum induction furnace or in electric furnace, and cast into from 20 kg to 1000 kg weigh ingots. Hot rolling equipment produces the 15mm thick plate specimens and hot extrusion machine forms the seamless tube specimens with from 8.1 to 9.0mm wall thickness and from 45 to 50.8mm outer diameter. Specimens were austenitized at from 1070 to 1100°C and quenched in air, then tempered at from 770 to 800°C.

4. Mechanical properties and microstructures

4.1. Microstructure

The optical micro-graph of NF12 is shown in Fig.2. The microstructure is full martensite phase without delta ferrite phase. The transmission electron micro-graph is shown in Fig.3. The carbides and nitrides precipitate on the primary austenitic grain boundary, on the martensitic lath boundary and also inside of the martensitic lath. We identified these carbides and nitrides as $M_{23}C_6$ and (Nb,V)(C,N) through X-ray diffraction qualitative analysis.

4.2. Mechanical properties

The experimental procedure and the dimensional details of test specimens are explained as follows.

The high temperature tensile test specimen with 30mm gage length and 6mm diameter reveals the tensile properties from at room temperature to at 700°C. The creep specimens with 30mm gage length and the 6mm diameter evaluate the creep properties at 600, 650, 700 and 750°C.

Charpy impact test specimen, full size or a sub-size of ASTM standard type with 2mm V-notch, prove the toughness at from -60°C to 60°C.

The result of high temperature tensile test is shown in Fig.4. 0.2% proof stress and tensile strength of NF12 at room temperature is over 500 MPa and 700 MPa respectively. These properties are approximately 1.4 times as high as those of Mod.9Cr-1Mo steel. The elongation and the reduction area ratio at room temperature are 18% and 70% or more respectively.

Creep rupture property of NF12 is shown in Fig.5. Fig.6 shows the extrapolated creep rupture strength of NF12 compared with those of NF616 and the Mod.9Cr-1Mo steel through the Larson-Miller parameter method. The extrapolated creep rupture strength for NF616(ASME T92/P92) is about 1.3 times as high as that of Mod.9Cr-1Mo steel (ASME T91/P91), and therefore NF616 is an epoch-making material compared with the conventional ferritic creep resistant materials. The creep rupture strength of NF12 is expected to be about from 1.3 to 1.5 times as high as that of NF616 depending of extrapolation method.

This result infers that the creep rupture strength of NF12 is supposedly about 1.6 and sometimes 1.9 times as high as that of Mod.9Cr-1Mo steel.

Fig.7 indicates the Charpy impact toughness of NF12. The absorbed energy at 0°C is higher than 100J/cm^2. NF616 and Mod.9Cr-1Mo steel are a little more tough, but the impact value for NF12 is still high enough to the application of martensitic creep resistant material.

High temperature tensile test, creep test and creep rupture test reveal the allowable stresses[6] at various temperature comparing with the allowable stresses for NF616 and Mod.9Cr-1Mo steel in Fig.8.

The allowable stresses for NF616 and NF12 tend to be much higher compared with that of Mod.9Cr-1Mo steel at relative high temperature. At 600°C, the allowable stress for NF616 is 1.3 times as high as that of Mod.9Cr-1Mo steel, and the stress for NF12 is 1.6 times high. The application temperatures are 596°C, 621°C and 645°C for Mod.9Cr-1Mo steel, NF616 and NF12 respectively when the allowable stress is determined to 70 MPa because of the difference of the high temperature tensile properties and creep properties for each steel.

Based on this calculated allowable stress, Fig.9 shows the comparison of the wall thickness of the presumably designed boiler tube for each steel when the steam pressure is assumed to be 27.5MPa and metal temperature is assumed to be 600°C and the inner diameter of the tube is assumed to be 40mm.

The estimated wall thickness for Mod.9Cr-1Mo steel is 11.5mm. However that for NF616 is 8.3mm and it results 36% reduction of cross section, and in case of NF12 51% reduction of cross section realizes 6.6mm wall thickness. In such severe steam conditioned fossil power plant, the cost for the welding and the fabrication increase because of the heavy wall thickness when Mod.9Cr-1Mo steel is applied. From these point of view, NF616 and NF12 have the big advantages on the fabrication cost over the conventional ferritic materials.

This estimation means that the application of NF12 realizes not only the very severe steam condition for the Ultra Super Critical power plant, but also the remarkable reduction of fabrication cost for the boiler construction.

5. Discussion

They say that dislocation movement controls the creep deformation mechanism in long time exposed ferritic steel at elevated temperature. Therefore introduction of obstacles against the dislocation movement is supposedly the most effective factor for the improvement of creep properties. The strengthening mechanisms of tungsten containing ferritic steels are expected through the two type of hypotheses, 1) increase of the creep deformation resistance by the inter-metallic compound precipitation to the dislocation movement on the glide plane or to the recovery of the microstructure during the creep when the main nucleation site of inter-metallic compound is lath boundary, and 2) increase of the resistance to the dislocation movement owing to the difference of the atom radius between tungsten and iron in the steel. The latter mechanism is explained as one kind of the solid-solution strengthening effects. Here, we discuss the two hypotheses.

Fig.10 indicates the precipitation amount of tungsten of NF616 mainly as a part of the inter-metallic compound during the creep exposure around at 600°C[7]. Tungsten precipitates only 0.2% at very beginning of the creep exposure as a metallic constituent of $M_{23}C_6$ type carbide. Then precipitation amount increases with exposure time and reaches about 1.3% after 100,000 hours according to the thermodynamic estimation. Therefore, the solution limit of tungsten can be estimated about 0.5% in NF616 steel at 600°C, and the almost excess tungsten precipitates as a constituent of inter-metallic compound (it is identified as Laves phase according to the precise analysis). If tungsten is contained beyond 0.5% in 9% chromium steel, excess tungsten precipitates as Laves phase, therefore the tungsten precipitation amount increases with the total tungsten content. The same precipitation behavior of tungsten was observed also in NF12. It infers the close relation between the creep property of about 2%

tungsten containing from 9% to 12% chromium steels and the decrease of the solved tungsten contents in those steels.

Not only carbides but also Laves phase coagulate in early stage of creep exposure. Additionally, the microstructure of inter-metallic compound is incoherent to the matrix. Therefore it is difficult to say that the Laves phase contributes to the creep rupture strength increase through the precipitation strengthening effect, and it can be rather concluded that the decrease of solution amount of tungsten with exposure time explains the deterioration of the creep properties judging from the concept above mentioned.

0.5% solved tungsten in 9 to 12% chromium steels possibly contributes to the stability of creep properties. Such solution strengthening effects can be explained by the resistance increase 1) to the dislocation movement on the glide plane through substitutionally solved tungsten dragging effect, and 2) to the vacancy diffusion around the dislocation core also through the substitutional solution of tungsten.

Two mechanisms are successive dislocation movements. It is difficult to explain the latter hypothesis by the calculated Miss-fit-parameter of tungsten, and only the precise analysis of Cottorel atmosphere of tungsten atom at around the dislocation core through the high resolution Transmission Electron Microscope is supposedly going to clarify the former hypothesis.

6. Conclusion

Successive effort of the research on NF616 and NF12 developed the tungsten containing martensitic creep resistant steels. Especially, the extrapolated creep rupture strength at 600°C for 100,000 hours is about from 1.6 to 1.9 times as high as Mod.9Cr-1Mo steel. NF12 is the most expected martensitic creep resistant steel for Ultra Super Critical steam conditioned fossil power plant application in the future.

References

1. T.Fijita, Current progress in advanced high Cr ferritic steels for high-temperature applications. Proc. COST-EPRI Workshop, Creep-resistant 9-12 Cr steels, Schaffhausen, Switzerland, 13-14 October 1986
2. H.Masumoto, M.Sakakibara, T.Takahashi and T.Fijita, Development of a 9%Cr-Mo-W steel for boiler tube, Proc. EPRI 1st int. conf. on improved coal-fired power plants, Palo Alto, USA, 20 November 1986
3. H.Masumoto, T.Takahashi and T.Fijita, Development and application of a 9Cr-0.5Mo-1.8W steel for boiler tube and piping. Proc. EPRI 2nd int. conf. on improved coal-fired power plants, Palo Alto, USA, 2 November 1988
4. H.Mimura, M.Ohgami, H.Naoi and T.Fujita, Development of 9Cr-0.5Mo-1.8W-V-Nb steel for boiler tube and pipe. Proc. High temperature materials for power engineering 1990 I , Liege, Belgium, 24-27 September 1990, pp.485-494
5. M.Ohgami, H.Mimura, H.Naoi and T.Fujita, Creep rupture properties and microstructures of a new ferritic tungsten containing steel. Proc. Creep characterization, damage and life assessment, Lake Buena Vista, Florida, USA, 18-21 May 1992, pp.69-73
6. ASME Boiler and pressure vessel code 1992, Section II Materials, Part D - Properties Appendix 1, pp.751-752
7. H.Mimura, M.Ohgami, H.Naoi and T.Fujita, Physical and mechanical properties of newly developed 9Cr-1.8W ferritic steel. Proc. TMS fall meeting, Pittsburgh

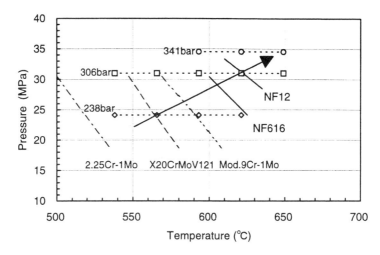

Fig. 1 Ferritic steels for Ultra Super Critical steam condition for fossil power plants.

Table 1 Chemical compositions of heat resistant steels.

(mass%)

Steel	C	Si	Mn	Cr	Mo	W	Nb	V	Ni	Co	B	N
NF12	0.08	0.20	0.50	11.00	0.15	2.60	0.07	0.20	<0.6	2.50	0.004	0.050
NF616	0.10	0.25	0.50	9.00	0.50	1.80	0.07	0.20	<0.4	-	0.002	0.050
Mod.9Cr-1Mo	0.10	0.35	0.50	9.00	1.00	-	0.05	0.20	<0.4	-	-	0.040
X20CrMoV121	0.20	0.25	0.50	11.50	1.00	-	-	0.30	0.55	-	-	-

Table 2 Chemical compositions of the NF12 steel.

(mass%)

Sign	Pruduct	C	Si	Mn	Cr	Mo	W	Nb	V	Ni	Co	B	N
NST	Tube	0.092	0.03	0.51	10.79	0.13	2.47	0.064	0.20	0.51	1.95	0.002	0.046
NDBT	Tube	0.094	0.19	0.50	11.35	0.16	2.57	0.070	0.20	0.41	1.49	0.004	0.040
TBE2	Plate	0.077	0.26	0.57	12.11	0.15	2.75	0.071	0.23	0.58	2.68	0.003	0.050
TBF2	Plate	0.078	0.22	0.47	10.85	0.14	2.60	0.068	0.20	0.51	1.91	0.003	0.048

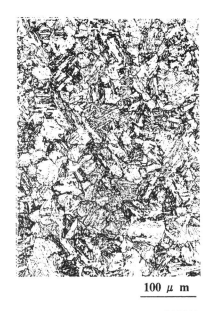

Fig. 2 Optical micro-graph of NF12.

Fig. 3 Transmission electron micro-graph of NF12.

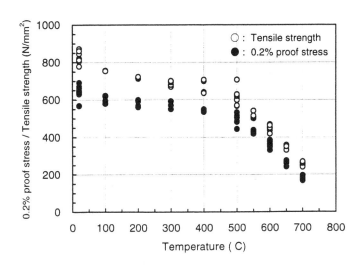

Fig. 4 High temperature tensile properties of NF12.

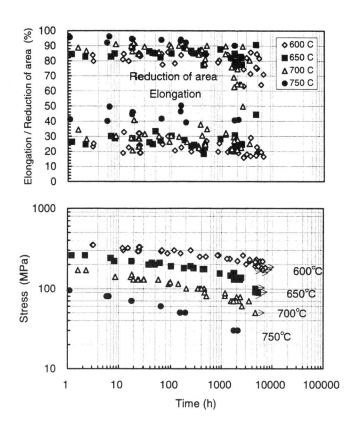

Fig. 5 Creep rupture properties of NF12.

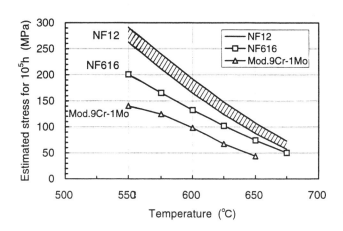

Fig. 6 Extrapolated creep rupture strength of heat resistant steels.

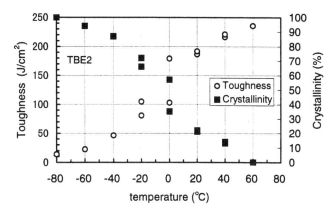

Fig. 7 Charpy impact properties of NF12.

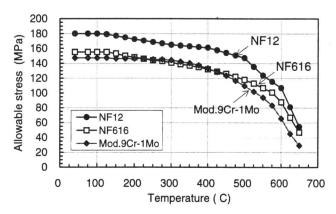

Fig. 8 Comparison of allowable stresses of heat resistant steels.]

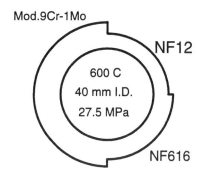

Fig. 9 Comparison of wall thicknesses of heat resistant steels.

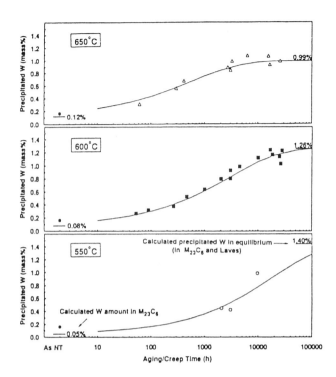

Fig. 10 Precipitation amount of tungsten of NF616.

C522/004/97

Experience in the manufacture of steam turbine components in advanced 9–12% chromium steels

M TAYLOR BSc, PhD and **D V THORNTON** CEng, FIM
GEC ALSTHOM Steam Turbine Group, Rugby, UK

SYNOPSIS

The award of contracts to GEC ALSTHOM Steam Turbine Group for steam turbines operating under advanced steam conditions has generated the need to use advanced 9-12% Chromium steels. The components purchased in the advanced materials have been in the form of rotor forgings, forgings for pipework and steam connections, forgings for diaphragm fabrication and castings.

For all types of forgings important parameters are steelmaking practice and heat treatment to achieve satisfactory mechanical properties. For rotor forgings the achievement of satisfactory centre-line properties in these large diameter components has been demonstrated by the testing of forged pieces of sizes representative of the rotor forgings and subjected to production heat treatments. The protection of the bearing surfaces of rotor forgings against deterioration during service has been achieved by overlay deposition.

Cast components produced in modified 9%CrMo steel have included inner casings, steam chest bodies and covers and cylinder internal thermal shields. Important production parameters identified have included steelmaking practice, mould design, heat treatment and the qualification of weld repair procedures. The need to achieve satisfactory properties in the weld repairs has generated specific compositional requirements for the filler metal. Manufacture has revealed variable weld repair rates which are generally consistent with the products being manufactured.

Manufacture of all components, whether in the form of large or small forgings, fabrications or castings has been satisfactory but it has been found necessary to pay attention to standardisation to achieve the required control parameters via satisfactory Quality Control at the manufacturing unit.

1. INTRODUCTION

The need to improve the efficiency is ever present in the development of new steam turbines. The most direct way of improving the efficiency of steam cycles is by increasing the steam temperature and pressure, but as a consequence changes to the materials used for certain components have had to be considered. The low alloy steels conventionally used for these components are restricted to temperatures of 540/565°C. Recently, advanced ferritic steels containing 9-12% Cr have been developed which can be used at temperatures up to 600°C.(1,2). This paper describes the technical aspects of the procurement of the steam turbine components in these advanced materials and the controls necessary to ensure satisfactory processing of these components.

Two 412MW double reheat steam turbines have recently been built for Skaerbaek and Nordjylland power stations in Denmark. The advanced steam conditions of 285 bar/580°C/580°C/580°C necessitated the use of advanced 9-12%Cr steels for certain high temperature components. These included rotor forgings, forgings for steam connections, general forgings, bar and rolled rings for the construction of fabricated diaphragms and numerous castings. Figures 1 and 2 show the components in the HP/IP cylinder and the VHP valve manufactured from 9-12%Cr steels.

2. MATERIAL SELECTION

The materials developments for application in steam turbines for fossil fired plant have recently been reviewed (3) and the specific development of new materials for advanced steam turbines is the subject of a paper to this conference (1). The latter development has largely resulted from the European collaborative development activity under the COST 501 programme which has been established and very well supported for about 12 years. The partners in the COST programme represent forgemasters, steel foundries, turbine manufacturers, boiler manufacturers, utilities and other users, testing institutes and universities. The aim of the COST programme was not only to develop materials having the required properties but also to demonstrate that these properties could be achieved in forgings and castings of sizes typical of the components used in advanced steam plant, and to establish a high temperature properties database for these materials. Similar developments have taken place in Japan and have resulted in similar alloys which have also been tested to demonstrate that the improved properties can be achieved in full scale components.

2.1 Rotor forgings

The COST programme identified a number of materials which exhibited satisfactory properties in the alloy development stage and which showed potential for the manufacture in the large pieces required for rotor forgings. A number of forgings were manufactured of diameters and weights typical of intermediate pressure (1200mm diameter) rotor forgings. The nature of these alloys with additions of Mo, V, W, Nb, N and B can lead to problems resulting from solidification of the large ingots required to produce rotor forgings. Therefore an important phase of the test programme was to destructively test the forgings to demonstrate:
- the achievement of satisfactory properties in the centre of the large forging sections
- satisfactory ultrasonic inspection characteristics

From the different steel compositions tested three steels achieved the target properties (4) and in particular provided a creep rupture stress in 10^5 hours which has a 50°C advantage over 1%CrMoV steels (Figure 3). Only two of these steels, grades E and F, had been proven in full size rotor forging production trials.

From an assessment of the results GEC ALSTHOM adopted the simpler, non tungsten bearing steel (grade F) 10%CrMoVNbN for the VHP and HP/IP rotor forgings of the Skaerbaek and Nordjylland machines.

2.2 Castings

The first of the new 9-12%Cr steels to be established in the early 1980s was the modified 9%CrMo steel developed by the Oak Ridge National Laboratories as a pipe steel, now ASME qualified as steel Grade 91. This was pursued as a casting alloy in an EPRI sponsored programme (5) and also was the subject of casting and fabrication trials in a collaborative CEGB and UK turbine makers programme (6).

Advanced castings were also being studied in the COST 501 programme; further modifications of Grade 91 either with higher carbon or a 1% tungsten addition, heat treated to a higher strength level, were investigated as laboratory trial melts and as a five tonne valve chest casting with a 1%W addition (7).

Throughout these trials this family of modified 9%Cr steels has been shown to have good castability, to be capable of through hardening in thick sections and to be weldable. The alloys investigated in COST 501 with 1%W showed an advantage in short term rupture strength as a result of the heat treatment to a higher proof strength, but at this stage the long term picture is not clear (Figure 4). The significantly easier weldability of the classic modified 9%CrMo steel and the adequately enhanced high temperature long term rupture strength led to this alloy being chosen for the high temperature castings of Skaerbaek and Nordjylland.

Founders require qualified welding procedures for the welding of castings during the up-grading operations and for construction welding of the attachments (inlet and exhaust steam connections). For welding of the modified 9%CrMo material a matching electrode is required. Although the material is easily weldable some of the electrode compositions available had reduced levels of the elements designed to impart good creep properties in order to provide acceptable impact properties. Therefore a GEC ALSTHOM filler metal specification was developed to ensure consistency of the welding electrodes used for casting repair and construction welding and to safeguard creep properties in all the welds.

2.3 General forgings, rolled rings, bar and plate

For these components ASME steel grade P91 was chosen on the basis of pre-existing technical data available from manufacturers for the section size of interest for these items.

3. ASSESSMENT OF SUPPLIERS

Before the placement of any orders for components in the new 9-12% Cr steels it was necessary to assess which suppliers were potentially capable of manufacturing the relevant products in these materials. At this time in early 1993 there was very limited production experience of large forgings, castings and general bar and forged product in these new steels.

For critical components, such as rotors and pressure containing castings, GEC ALSTHOM require suppliers to be qualified for manufacture. Qualification can be obtained by

demonstrating previous experience in the manufacture of the component in the required material. Where no such experience exists, alternatives for qualification are available as follows:
- previous experience of manufacture of a similar component in the required material
- previous experience of manufacture of the required component in a similar material
- previous experience of supply to GEC ALSTHOM of the required component
- demonstrated technical expertise

In the latter case it is possible for the placement of an order to be conditional on the supplier carrying out qualification testing in parallel with the manufacture of the component.

On the basis of responses to questionnaires, a short list of suppliers was drawn up, a number of whom were identified for further technical discussions.

3.1 Rotor forgings

The production rotor forgings were required without a centre bore. Therefore for suppliers, who had not previously manufactured 9-12%CrMo rotor forgings, through boring of the contract rotor forgings during manufacture was not available as a method of qualification. Alternatives for verifying satisfactory properties are:
- forging of a test piece of equivalent dimensions to the contract rotor forging but separate from it.
- forging of a test piece of equivalent dimensions to the contract rotor forging but integral with it.
- forging of an overlength of the body diameter of the contract rotor forging and subsequent testing of the centre-line properties under this extension, the excess material being removed during the machining of the shaft end profile.

The use of material with greater than 3% Chromium for journal bearings in steam turbine rotor forgings leads to a wear phenomenon whereby the surface deteriorates during service and debris is produced which resembles wire wool. This problem can be overcome by replacing the bearing surface with material containing less than the critical chromium level. A number of processes have been employed to achieve this protection such as plasma spraying, weld overlay, chromium plating and mechanical sleeving. For the Skaerbaek and Nordjylland rotors mechanical sleeving was not an option. The weld overlay method was chosen as a result of a collaborative development programme between GEC ALSTHOM, Parsons Turbine Generators and Forgemasters Engineering and Steels Limited which offers a robust long term solution to this problem. The welding procedure to be employed during the manufacture of the rotor forgings formed a significant part of the technical discussions held with the potential suppliers.

On completion of the technical discussions, a number of suppliers were identified who were already technically qualified for manufacture or who were prepared to undertake a qualification using one of the above methods in parallel with the rotor forging manufacture including the requirement for the weld overlay of the bearing surfaces.

3.2 Castings

The questionnaire responses indicated that a number of foundries did have limited experience with the modified 9%CrMo castings. Suppliers not qualified need to be able to demonstrate an understanding of the differences required in basic foundry practice, such as mould design (contraction allowances) and sand practice, when dealing with high chromium steels as compared with low alloy or plain carbon steels. Several suppliers had experience of manufacture of other grades of high chromium steels and were therefore aware of potential problems. Of those suppliers with no prior experience but with the capability of producing the larger items, a number manufactured test castings in order to establish heat treatment parameters, verify achievement of specification properties in the large sections and carry out a weld procedure qualification.

A large number of the castings needed were of weights less than two tonnes. As the majority of foundries under consideration for these items would require to qualify weld procedures, the cost of such qualification would represent a significant proportion of the cost of the castings Therefore suppliers of the larger castings were encouraged to also manufacture smaller castings, the results of the qualification testing carried out for the larger castings being applied to the smaller and thereby improving the economics of small modified 9%CrMo castings. Similarly, large numbers of the smaller castings were grouped together to make an attractive technical and commercial package in order to offset qualification costs.

4. MATERIALS PROCUREMENT

4.1 Rotor forgings

Orders were placed with a UK forgemaster for the two HP/IP rotors and a Japanese and an Austrian forgemaster for the VHP rotors, a total of four rotors being required for the contract. The UK forgemaster was a partner in the COST development programme and therefore, with the exception of a qualification for the weld overlay procedure, no further qualification testing associated with the contract rotor forgings was required. The Japanese and the Austrian forgemasters opted to forge an integral qualification test piece on the end of the rotor forging which was of equivalent dimensions to the rotor forging Itself. This was used for qualification of the manufacturing procedure by destructive testing to ascertain the mechanical properties at different positions within the forging and also to qualify the weld overlay procedure prior to it being applied to the bearing surfaces at a later stage in the manufacture

Important parameters to be considered in the manufacture of rotor forgings, irrespective of material, are:

- steelmaking and ingot design
- forging control
- heat treatment
- non-destructive examination (NDE)

Because of the high chromium content of the material it was important to control the steelmaking process so as to avoid the oxidation of the material and hence defects within the final forging. Pouring of the ingot was carried out under an Argon shield to avoid re-oxidation of the steel. The high chromium and other alloy content of the steel also leads to adverse segregation during ingot solidification which can give unsatisfactory properties in the final forging. Methods of minimising this are by suitable ingot design and by using sophisticated ingot hot topping procedures, such as Electro-Slag Remelting, to ensure adequate feeding and solidification of the ingot. Both these methods were used by the forgemasters during manufacture of the rotor forgings.

It has long been established that cracking of 9-12% Cr steel can occur during forging if excessive strains are applied during any of the forging operations. It was therefore necessary to increase the number of forging heats in order to apply the total deformation required for consolidation and structural control over an increased number of forging operations, which allows the material to 'recover' and hence accommodate further strain during the subsequent forging cycle.

Rotor forgings are subjected to high sensitivity non-destructive examination (NDE) to ensure that any defect remaining in the forging will not exhibit significant extension during the service operation of the rotor. The ability of ultrasonics to detect defects lying on the forging axis at the sizes required depends on the structure of the material. Steels containing high chromium contents are less transparent to ultrasound than those made of low alloy steels, the ability of ultrasound to penetrate decreasing with increasing thickness. It is important therefore that the material structure is optimised to enable ultrasonic examination to be applied. Heat treatment is an important process in controlling the structure, not only the quality heat treatment which generates the mechanical properties but also the annealing applied after forging which 'conditions' the structure so that it responds to the quality heat treatment. The GEC ALSTHOM specification requires defects of 1.6mm Flat Bottom Hole Equivalent (FBHE) response to be detectable throughout the forging. This sensitivity was achieved on all forgings and, in the case of one of the VHP rotors having a maximum diameter of 770mm, a minimum detectability of 1mm FBHE was achieved. (Table 1).

Table 1 Ultrasonic detectability for 10CrMoVNbN rotor forgings

Rotor type	Test diameter (mm)	MDDS (FBHE) (mm)
VHP #1	790	0.8
VHP #2	795	1.1
HP/IP #1	1200	1.3
HP/IP #2	1200	1.4
Requirement		≤ 1.6

MDDS: Minimum detectable defect size on the axis (Flat bottom hole equivalent diameter)

Table 2 Summary of properties for 10CrMoVNbN rotor forgings

Rotor Type	Weight (t)	Diameter (mm)	Test Location	$R_{p0.2}$ (MPa)	R_m (MPa)	Z (%)	FATT (°C)
VHP	10.7	780	Exterior	734/795	869/893	51/55	+20/+45
			Axis	738/795	888/917	53/56	+10/+52
HP/IP	20	1180	Exterior	731/817	872/948	48/56	+48/+54
			Axis	729/751	855/895	51/52	+15/+40
Requirement				700 min	800-950	40 min	+60°C max

Heat treatment: 1070°C/570°C/675°C

Table 2 shows the actual mechanical properties achieved on the rotor forgings made during qualification and for the contract. The rotors show a satisfactory hardenability with the strength on the axis being similar to that on the surface. The FATT values indicated a reasonable margin against the specification level of +60°C demonstrating that the toughness of the more highly creep resistant 10%CrMoVNbN steel is superior to that of the traditional 1%CrMoV alloy which has an FATT of +120°C.

The weld overlay of the bearing surfaces was carried out by all forgemasters using the submerged arc technique. Because of the nature of the welding process, there was the probability that defects could be introduced into the rotor forging which were not normal for a forged product, viz. minor welding defects in the direction of welding which lay in a transverse direction in the rotor forging. The service conditions demanded that such defects, although small, should be capable of detection so that their acceptability could be established. Ultrasonic examination of the bearing areas was carried out using the shear wave technique as routinely applied to the examination of high integrity welds such as steam chest butt welds. Some of the weld overlays did contain defects which were assessed as unacceptable and these had to be removed and the resulting excavation repaired by a further welding operation.

The thermal expansion coefficients of the weld overlay (<2%Cr) and the base material of the rotor forging (10%CrMoVNbN) are sufficiently different to give rise to residual tensile stresses in the weld overlay. After completing the overlay the welded area is heat treated at 620°-640°C to provide tempering of the structure and to relieve welding stresses. However the differential expansion of the weld deposit and the rotor between the stress relieving temperature and ambient still resulted in residual tensile stresses in the low alloy steel weld overlay. This effect can be nullified by cold rolling of the bearing surface to promote residual compressive stress throughout the weld overlay and eliminate any potential reduction in fatigue resistance caused by high tensile mean stress.

All rotors were manufactured satisfactorily and delivered to GEC ALSTHOM Rugby works in a fully compliant and acceptable condition. The fully bladed HP/IP rotor is shown in Figure 5.

4.2 Castings

Forty critical castings and more than forty general castings in modified 9%CrMo steel were required for the Skaerbaek and Nordjylland contracts and orders were placed for the castings with foundries in the United Kingdom, Italy, France, Austria and Germany.

Important parameters in the manufacture of castings are:
- steelmaking
- mould design and sand practice
- establishment of weld procedures
- weld repair rates

For the modified 9%CrMo steel it was necessary that the steelmaking procedure addressed achievement of low sulphur contents to minimise hot tearing during solidification and, as the analysis required the addition of nitrogen, the need to ensure that this was added at the correct time in the steelmaking so that gas evolution during the solidification in the mould was avoided.

Moulding contraction allowances had to be established to take account of the thermal expansion characteristics of the material. Sand practice involved the use of refractory chromite sand and a mould dressing of zirconia paint for those parts of the mould subject to liquid metal contact to prevent the attack by the more aggressive liquid 9%CrMo steel.

No significant problems with the achievement of mechanical properties were anticipated as the test castings produced had shown that satisfactory properties could be achieved even in the thickest sections of the largest castings by forced air cooling from the hardening temperature. This was reflected in the mechanical properties achieved as seen in Table 3. The toughness was quite variable as may be expected with castings where a significant variation in grain size and structure may be encountered. This led to the selection of the highest possible tempering temperature to improve toughness compatible with the achievement of the tensile properties.

Table 3 Summary of properties of modified 9CrMo castings

Component group	Weight (t)	$R_{p0.2}$ (MPa)	R_m (MPa)	Z (%)	Cv (mean) (J)
Inner casings	9-15	503/545	648/707	58/66	46/78
Steam chest castings	4-16	500/574	624/730	28/68	123/31
Cylinder internal castings	0.5-2	507/591	645/731	16/23	33/64
Requirement		500 min	600-750	15 min	30 min

Heat treatment: 1050°C/740°C/730°C

All foundries had to undertake qualification of welding procedures for the specified casting material and filler metal to meet the quality requirements for both weld repair and the construction welding of attachment forgings. No difficulties were experienced by the foundries in achieving the property requirements in the weldments following the post weld heat treatment at about 740°C.

Conventional NDE techniques (Magnetic Particle Inspection and ultrasonic examination) as applied to low alloy steam turbine castings were used to examine the castings for defects. Ultrasonic examination employed compression and shear wave probe techniques to detect and evaluate shrinkage and tear-like defects. No differences in ultrasonic detectability to that achieved in low alloy steel castings was noted. Unacceptable defects were removed and the resulting excavation subjected to weld repair.

Table 4 Weld repair rates

Component group	Weight range (t)	Average repair rate (%)
Inner casings	9-15	0.5
Steam chest castings	2-16	1.2
Cylinder internal castings	0.5-2	0.5

Weld repair rates for modified 9%CrMo steel are shown in the Table 4. Some modest increase in repair rate compared with low alloy steel was observed in castings from foundries with only limited previous experience of the modified 9%CrMo steel. However taking into consideration the complex shape of some of the castings, particularly the steam chest bodies (Figure 6), the weld repair rates are consistent with the products being manufactured.

4.3 General forgings, rolled rings, bar and plate

Material purchased in these forms was used as follows:
- general forgings - steam connections
- general forgings - valve internals (spindles, seats etc.)
- rolled rings, bar and plate - diaphragm manufacture

These forgings were sourced in a number of countries throughout Europe including the UK, Italy, France. The modified 9%CrMo material chosen for these items was becoming more widely used as the alloy had been ASME qualified and was being generally applied as a pipe and tube material. No particular problems were experienced by suppliers in obtaining the material as ingot or semi-forged product. The alloy caused no processing problems either in hot working (forging, rolling and ring rolling) or heat treatment.

These bought in materials were subjected to further processing either at GEC ALSTHOM works or by subcontractors for use in the turbine assembly to manufacture components such as steam connections, valve internals and diaphragms.

4.4 Steam connections

Steam connections were purchased by GEC ALSTHOM and supplied to those foundries involved in the manufacture of the steam chest castings and the outer cylinder castings for attachment to the castings by welding. The steam chests were manufactured in the modified 9%CrMo steel and therefore the welding procedures developed for the casting repair were also applicable to the welding of the same composition attachments.

However the outer cylinder castings were manufactured in low alloy steel material. The attachment of modified 9%CrMo material to low alloy steel presents a number of problems. Dissimilar alloy joints can develop problems which result from differences in mechanical properties, differences in heat treatment and the presence of compositional gradients leading to carbon migration. The development of a welding procedure for modified 9%CrMo to low alloy steel had previously been undertaken (8). However the logistics of applying a suitable overall heat treatment schedule had to be formulated for the steam connections to the outer casing.

The heat treatment requirements for the two materials were such that the post weld heat treatment temperature for the attachment weld of the 9%CrMo forging would lower the strength of the casting to below the design requirement. The solution to this was to introduce a 'transition piece' in low alloy steel between the 9%CrMo forging and the low alloy steel casting. The joint design is as shown in Figure 7.

The transition piece was first welded to the modified 9%CrMo forging using a 9%CrMo filler and the sub-assembly stress relieved in accordance with the requirements for the high alloy weldment. In order to prevent overtempering the transition piece was purchased in a condition such that the stress relieving temperature of the sub-assembly weld did not exceed the tempering temperature of the transition piece. The transition piece to casting weld was then made using a low alloy steel filler and the whole assembly stress relieved at a lower temperature appropriate to low alloy steel so that the casting strength was maintained and the weld fully stress relieved.

4.5 Valve components

The area of the seating line of the valve heads and valve seats and certain sliding surfaces of valve components are protected against surface deterioration and hence loss of performance. The components are machined from modified 9%CrMo bar and forgings and the cobalt hard facing protection is applied by weld deposition using plasma transferred arc or tungsten inert gas processes. The modified 9%CrMo material has been somewhat easier to hardface than the higher carbon 12%CrMoV previously used.

4.6 Fabricated diaphragms

GEC ALSTHOM commonly use a welded diaphragm construction which employs spacer bands to accurately locate the blades between the rim and centre forgings as shown in Figure 8. For operation at the higher temperatures and pressures in Skaerbaek and Nordjylland, modified 9%CrMo offers an excellent combination of creep strength and fabricability and, in addition, matching consumables with high creep strength were available. However, manufacture of trial diaphragms demonstrated that under conditions of high restraint and in the presence of notches, careful quality control with respect to welding parameters such as preheat, weld sequence and temperature differentials must be applied, even with materials which normally have excellent weldability, in order to avoid cold cracking. For the thickest diaphragms an additional precaution was to introduce an intermediate stress relief for welds above 60mm deep.

5. CONCLUSIONS

The experience gained by GEC ALSTHOM and suppliers in the manufacture and processing of components in advanced 9-12%CrMo steels has demonstrated that, provided the differences to low alloy steels in processing are recognised and appropriate manufacturing procedures formulated, large pieces in these materials can be produced with the required properties. In addition, these materials do not present any significant manufacturing problems but are noted to require a high standard of technical control at all stages of the manufacturing process.

The manufacture of a variety of components including large forgings, castings and fabrications in the new 9-12% chromium steels provides a firm foundation for the future in which more highly alloyed steels of this family will be required as the inlet steam temperature and pressure are increased as a means of increasing the efficiency of large steam turbines.

References

1. Thornton, D.V., Vanstone, R.W. New materials for advanced steam turbines, this conference.

2. Paterson, A.N. Steam turbines for advanced steam cycles, this conference.

3. Thornton, D.V., Vanstone, R.W. Materials development for application in steam turbines for fossil fired plant, Materials Engineering in Turbines and Compressors, Newcastle upon Tyne, April 1995.

4. Berger, C., Scarlin, R.B., Mayer, K.H., Thornton, D.V., Beech, S.M. Steam turbine materials: High temperature forgings, Materials for Advanced Power Engineering, Liège, October 1994.

5. Mayer, K.H., Gysel, W. Modified 9%CrMo cast steel for casings in improved coal-fired power plants, Proc. of 3rd EPRI Int. Conf. on Improved Coal-Fired Power Plants (ICPP), San Francisco, April 1991.

6. Thornton, D.V., Hill, R. The fabrication and properties of high temperature high strength steel castings, ibid.

7. Scarlin, R.B., Berger, C., Mayer, K.H., Thornton, D.V., Beech, S.M. Steam turbine materials: High temperature castings, Proc. of COST 501 Conf. Materials for Advanced Power Engineering, Liège, October 1994.

8. Thornton, D.V. Materials for modern high temperature steam turbines, Proc. of I. Mech. E. Conf. on Steam Turbines for the 1990s, London, 1990.

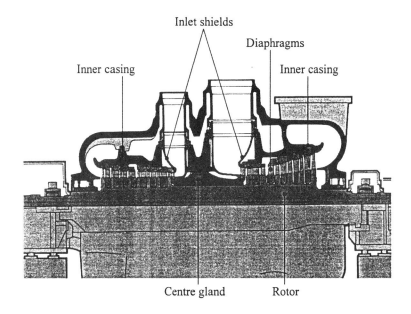

Fig. 1. HP/IP Cylinder showing major components in advanced 9-12%CrMo materials

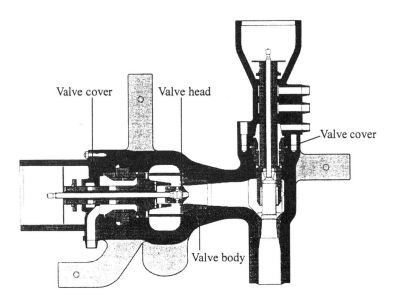

Fig. 2. VHP valve showing major components in advanced 9-12%CrMo materials

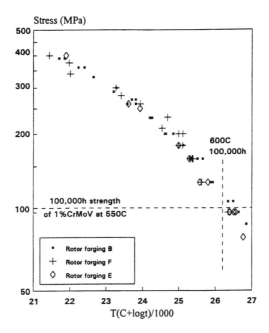

Fig. 3. Rupture strength of rotor forgings developed in COST 501

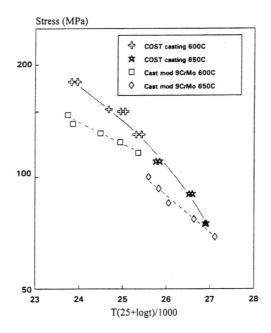

Fig. 4. Rupture strength of recently developed casting materials

Fig 5 Fully bladed HP/IP rotor

Fig 6 HP steam chest

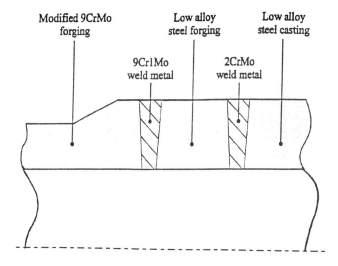

Fig. 7. Transition joint: low alloy steel casting to modified 9%CrMo steam connection

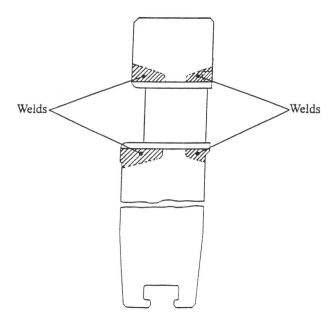

Fig. 8. Welded diaphragm construction

Boilers

C522/023/97

Benson boiler with vertical tube water walls – principles and advantages

J FRANKE, R KRAL, and E WITTCHOW
Siemens Aktiengesellschaft, Power Generation Group (KWU), Germany

1 Introduction

To date, the spiral arrangement of the evaporator tubes in the furnace walls has been typical for Benson boilers. This concept is well-tried in more than 300 boilers over the past 30 years. However, the pressures of competition have demanded reductions in manufacturing and installation costs, and there is a need to aid the introduction of Benson boilers in countries where drum boilers have so far been dominant. These challenges have led to further development of the vertical-tube furnace Benson boiler concept originally introduced in the 1950s (Fig. 1). The technical improvements are described in this paper.

2 Requirements for a Vertical-Tube Once-Through Evaporator

Specific design requirements must be met independent of the type of evaporator tubing. These are reliable tube cooling on the one hand and the most uniform possible temperatures at the outlet of all parallel evaporator tubes on the other. In the case of spiral-wound evaporator tubing, the inclination of the tubes can be selected to achieve nearly any mass velocity required for reliable cooling. In addition, the effects of nonuniform heat input are slight, as each tube passes through nearly all of the heat transfer zones in the furnace.

In vertical tubes geometric aspects result in a significant decrease in mass velocity. Cooling of the tubes can therefore only be ensured by supplementary measures to improve internal heat transfer. This requirement can be met by the use of internally-rifled tubing.

Heat absorption over the furnace circumference is non-uniform, with its distribution depending on the firing system selected. For example, in the case of an opposed firing system heat input to evaporator tubes in the furnace corners is considerably less than to those in the centre of the furnace walls. Therefore, in order to achieve approximately equal outlet temperatures, the furnace design must ensure that flow in the tubes with less heat input is at a lower mass velocity, while flow in those with more heat input is at a higher mass velocity. The Siemens concept meets this requirement by means of a design featuring low mean mass velocities.

3   Heat Transfer in Tubes

Heat transfer in the evaporator tubes depends on the wetting condition of the tube wall. This holds both for smooth tubes and for rifled tubes. When steam quality reaches a certain level, wetting of the tube wall can no longer be maintained. There is a sharp reduction in heat transfer on the unwetted tube wall. If heat flux remains unchanged, this results in a significant increase in inner tube wall temperatures. Fig. 2 gives a comparison of this behaviour for a smooth tube and for a rifled tube. In the rifled tube, wetting of the tube wall is maintained until the steam quality is > 0.9, shortly before evaporation is completed. This compares to a steam quality of 0.6 for a smooth tube. This can be explained by the angular momentum imparted to the flow by the spiral ribs. The differing centrifugal force results in a separation of the water fraction from the steam fraction. The water is pressed against the tube wall. Also at the end of the wetted region, there are high flow velocities due to the high steam quality level. This promotes good heat transfer even for the dry wall and hence only a smaller scale increase in tube wall temperature.

The investigations performed by Siemens show that sufficient cooling of the evaporator tubes is ensured for rifled tubes with feedwater pressures below 200 bar even for mass velocities of 150 kg/m^2.s. This permits operation at a part-load of 20% and below in once-through mode. However, as for smooth tubes, degradation of heat transfer cannot be avoided in the vicinity of critical pressure, generally between roughly 200 bar and critical pressure. As this can also occur in the burner region, i.e. in the region with high heat flux, the design must account for the resulting increase in wall temperatures.

A further step in fundamental research was the optimisation of the rifling pump geometry, which is characterised by such parameters as the shape and height of the ribs. Fig. 3 shows the inner wall temperatures for a typical commercial tube and for an optimised tube. Given the same boundary conditions, the mass velocity in the optimised tube can be significantly reduced relative to that for the commercial tube to achieve the same inner wall temperatures in the pressure range critical to heat transfer.

Siemens began fundamental research on tubes with rifling in the 80s. The parameter range of the investigations is shown in Fig. 4, where tubes RRA to RRH represent a wide range of diameters with various rib geometries. Siemens has since compiled the world's largest database in this field, with roughly 100,000 data points. This knowledge base corresponds in scope and degree of detail to information obtained on heat transfer and pressure drop in smooth tubes, thus ensuring the same degree of design safety for Benson boilers with rifled tubes as that provided by the validated knowledge base for spiral-wound Benson boilers with smooth tubes.

4   Once-Through Evaporator with Natural Circulation Characteristic

Spiral wound tubing in the furnace walls passes through areas of different heat flux, which tends to even out higher and lower heat input to single tubes to a great extent. Measurements of heat flux taken on a 700 MW coal-fired boiler indicated that increased heat input into a vertical tube relative to the mean heat input for all tubes can be up to three times greater than that for a spiral tube (i.e. 15% instead of 5%). A vertical tubing concept must therefore find a new approach to the problem of heat input variations. The basic principle of the Siemens concept is to make use of the natural circulation characteristic to adjust the mass-flow in a tube to its heat input. A natural circulation characteristic means that higher heat input to a single tube in an evaporator heating surface results in higher mass-flow for this tube (Fig. 5).

This concept is based on the assumption that good heat transfer can be achieved in tubes with rifling even at very low mass velocities. If mass velocity is appropriately low, this results in the following behaviour of a single tube with excess heat input in a parallel-tube system:

- The reduced weight of the water/steam column causes the hydrostatic pressure drop to fall with increased steam generation.

- At the same time, the frictional pressure losses rise due to the higher steam fraction and the associated increased flow velocity.

- As the magnitude of the change in the hydrostatic pressure drop is significantly greater than the change in frictional pressure losses, this would result in a decrease in overall pressure drop (Fig. 6).

- The reduced pressure drop, acting via the common inlet and outlet headers, then causes the mass flow through the single tube with excess heat input to increase to match it.

The flow characteristic of the evaporator system is thus comparable with that of a drum boiler, i.e. a natural circulation characteristic. This results in a self-compensating effect with regard to the consequences of nonuniform heat input in a vertical tube system.

5   Testing of the Concept in a Power Plant

It is not possible to construct low-output demonstration plants to test vertical-tube concepts, as the lower application limit is roughly 300 MWel for supercritical pulverised-coal-fired boilers due to thermohydraulic considerations, and roughly 250 MW for subcritical boilers of this type. Three evaporator tubes were therefore installed and operated for more than 10,000 h as test heating surfaces in the supercritical 320 MW Benson boiler in Farge power plant. The test rig was installed, with the approval of the utility PreussenElektra, by a consortium consisting of Babcock Lentjes Kraftwerkstechnik GmbH, Siemens Power Generation (KWU) and L&C Steinmüller GmbH. The test rig layout and instrumentation are shown in Fig. 7.

The test performed in Farge power plant served both for experimental verification of theoretical data on flow distribution in the event of heat input variations and for verification of system insensitivity to transients. It was important to perform the test with a number of features that are of major significance for the natural circulation characteristic, such as a furnace of real height, that could not be implemented in the laboratory.

Details of the test rig and the test procedure are given in /1/. The natural circulation characteristic predicted by computation was clearly verified under realistic conditions in Farge power plant. The tube with the highest heat input also had the highest mass-flow of all three tubes (Fig. 8). In addition, it was found that were no deposits on the inside of the tubes at the end of the 10,000 hour test duration.

6 Operating Advantages of the Advanced Concept

Benson boilers with vertical tubes have a number of additional operating advantages over Benson boilers with spiral-tube furnaces as follows:

- Reduction of evaporator mass flow to 20% BMCR or possibly even lower. The start-up circulating pump is then no longer required, this greatly simplifying the start-up system.

- Reduction of Benson minimum load to about 20% BMCR, but retaining high main steam temperatures. This means that nightly and weekend outages are not required in most cases.

- Evaporator pressure drop decreases by about 5 bar.

Computer simulation was performed to analyse start-up without a circulating pump and with an evaporator mass flow of 20%. A reference calculation was performed for the conventional type of Benson boiler (spiral-tube arrangement) with a minimum Benson load of 40% and a start-up system with a circulating pump. Data from experience were used as a basis for calculations for start-up status of the steam generator following a weekend outage. The boundary conditions assumed for combustion control and bypass station operating mode were the same in each case.

Fig. 9 shows the main steam flow and temperature and hence the behaviour of the steam generator for the various concepts. In both cases, roll-off conditions for the turbine in terms of temperature and main steam flow are achieved after roughly one hour. Both the steam flow as well as the increase in main steam temperatures is approximately equal for both concepts.

There are also no significant differences in terms of start-up heat losses. Compared to operation with use of a circulating pump, bypass to the flash tank involves higher flash losses, but correspondingly lower condenser losses.

In addition to the two concepts described, the figures show the behaviour of steam generation and main steam temperatures for a steam generator with a spiral-tube furnace, start-up without a circulating pump and an evaporator system mass flow of 40%. As is to be expected, the main steam flow is significantly smaller as a result of the lower evaporator inlet temperature, with associated disadvantages for cooling of the superheater heating surfaces.

A further operating advantage associated with start-up without a circulating pump is the resulting simplification of level control.

In addition, calculations indicate that evaporator outlet temperature is significantly less sensitive to transient processes such as load changes or burner changeover in the case of the vertical-tube configuration (Fig. 10). This would lead one to expect at least the same load change rates as are standard for Benson boilers with spiral-tube furnaces.

As has already been mentioned, a vertical-tube configuration with optimised rib geometry permits the minimum load in once-through operation to be reduced to 20% and possibly even lower from the previous 35 - 40% for spiral-wound furnaces with smooth tubes, but without decreasing the main steam temperature. The minimum load for coal firing is 20% in most cases. A reduction of the minimum load to this level eliminates the need for the transition from once-through operation to operation with circulating and vice versa, which would otherwise entail a hold point and also additional material stress loadings both on the boiler and at the turbine inlet due to temperature changes. Frequent start-up and shutdown can be avoided because the plant can now be operated at very low loads with high efficiencies. This represents a significant advantage for future plants with elevated conditions and thick-walled components which permit only slow start-up and shutdown.

Service advantages can also be anticipated due to the simple vertical tube configuration, for example the furnace walls will be less prone to fouling.

7 Advantages in Terms of Investment Costs

The Siemens´ concept with vertical tubes features lower investment costs than a spiral-tube furnace with smooth tubes. The furnace walls with vertical tubes enable simple and hence cost-effective construction as for a drum boiler. Additional welded tie bars as for spiral furnace tubes are not necessary, and the assembly welds in the furnace corners are simple longitudinal welds. Further major savings can be realised in the start-up system. The low minimum load of 20% permits implementation of the start-up system as a simple system with a bypass to an atmospheric flash tank. The start-up losses are so low due to the minimum evaporator flow and the early transition to once-through operation that circulating pumps, bypass heat exchangers or return flow to the feedwater tank are not necessary.

8 Summary

On the basis of extensive research into heat transfer and pressure loss in tubes with internal rifling, the Benson boiler with a vertical-tube furnace has been further developed by Siemens. Optimisation of the internally rifled tubes for good heat transfer permits a significant reduction in mass velocity in the tubes, thus allowing the hydraulic design of the furnace walls to produce increased mass flow in tubes with high heat input. This type of behaviour is familiar from natural circulation systems. The correctness of the hydraulic design was verified convincingly in a large-scale test on a supercritical 320 MW boiler.

The concept of vertical tubes combines the design advantages of the drum boiler with the operating advantages of a Benson boiler with spiral evaporator tubes, and also yields additional operating advantages.

This vertical-tube configuration can be used to realise a cost-effective boiler with outstanding operating characteristics for the unit size range from about 250 MW (with this lower application limit being far lower in the case of atmospheric fluidized bed boilers) to over 1000 MW.

References:

/1/ Franke, J., Cossmann, R., and Huschauer, H:
Benson Steam Generator with Vertically-Tubed Furnace, Large-Scale Test under Operating Conditions Demonstrates Safe Design, Published in VGB Kraftwerkstechnik 75 (1995), Number 4.

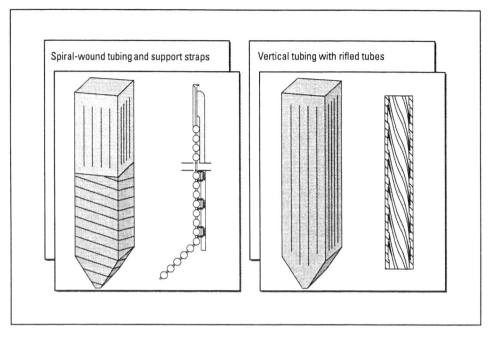

Fig 1 Water wall concepts for BENSON boilers

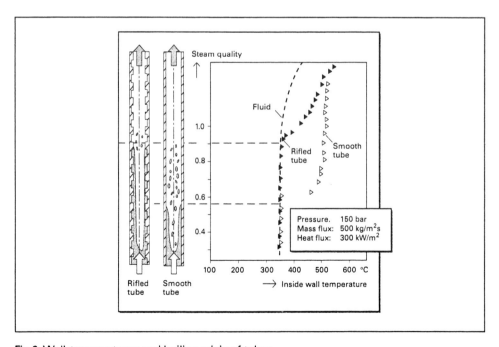

Fig 2 Wall temperatures and boiling crisis of tubes

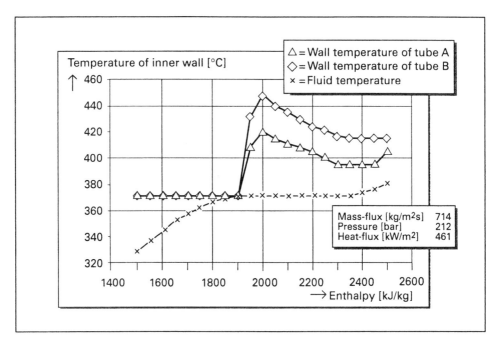

Fig 3 Heat transfer improvement potential through optimized rib geometry

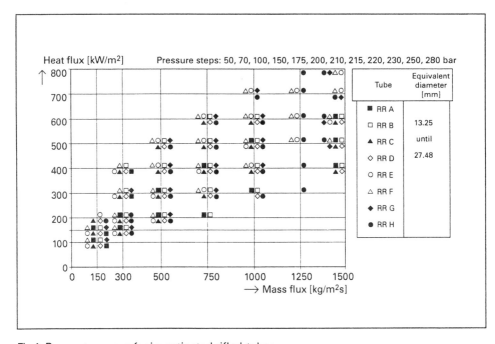

Fig 4 Parameter range for investigated rifled tubes

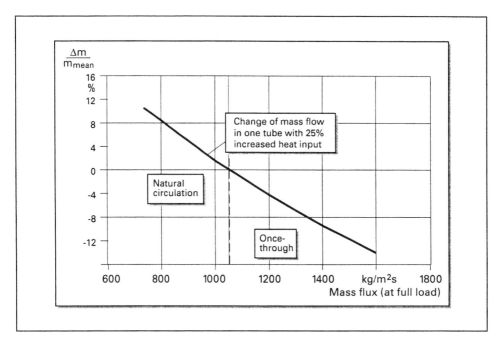

Fig 5 Mass flow characteristic of verticall tubed BENSON boilers

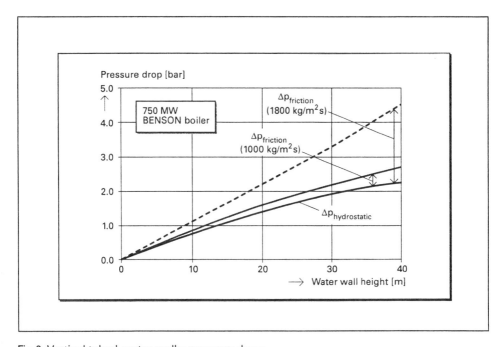

Fig 6 Vertical tubed water walls: pressure drops

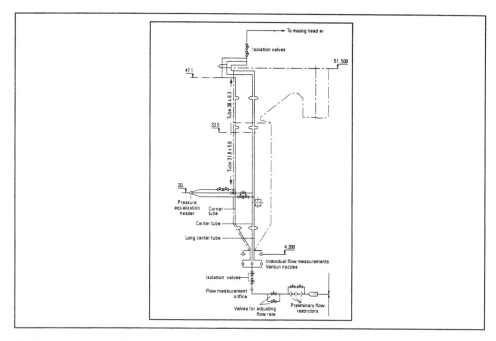

Fig 7 Farge power plant test rig

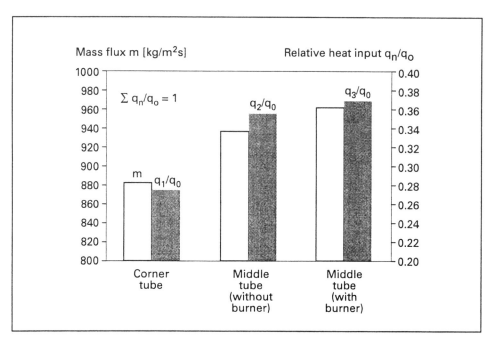

Fig 8 600 MW BENSON boiler with vertical tubed water walls
(Measurements)

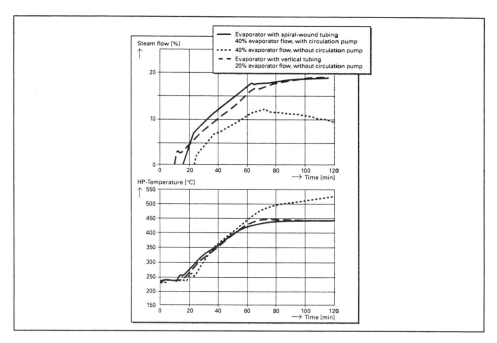

Fig 9 Steam flow and HP-temperatures during start-up after weekend shut down

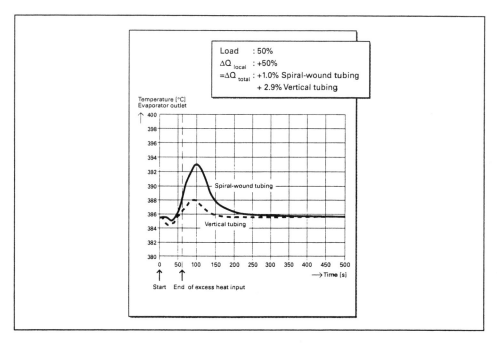

Fig 10 Temperature increase due to short-time, local excess heat input in the burner region

C522/023 © IMechE 1997 153

C522/005/97

The design of a 1000MW coal-fired boiler with the advanced steam conditions of 593°C/593°C

K SAKAI BSc, Thermal and Nuclear Power Technology Association and **S MORITA** ME, MJSME
Babcock-Hitachi K K Kure Works, Hiroshima, Japan

SYNOPSIS

Babcock-Hitachi K.K. Japan, has been constructing a 1000MW coal fired boiler with the advanced steam condition of 24.1MPa/593°C/593°C, which will be supplied to Matsuura Thermal Power Station No.2 Unit for Electric Power Development Co., Ltd., and will commence commercial operation in July, 1997. This paper presents the design features of this supercritical sliding pressure operation boiler especially designed for middle load usage. State-of-the-art low NOx combustion technology has been applied to this plant to meet tight emission control requirements. The performance of the boiler will be confirmed through the coming commissioning period.

1. INTRODUCTION

In 1990, Babcock-Hitachi K.K. (BHK) completed Matsuura Thermal Power Station No.1 Unit for Electric Power Development Co., Ltd. (EPDC), which was the first 1000MW coal fired boiler in Japan, with a steam condition of 24.1MPa/538°C/566°C. The operation record of this boiler has been satisfactory and the plant has been contributing to a stable supply of electricity in the western part of Japan (1).

Nowadays, Japanese utility companies are requested to reduce air pollutant emissions particularly CO_2 and steam conditions of thermal power plants have been improved drastically in Japan. EPDC has been paying special attention to environmental issues and decided to apply an advanced steam condition of 593°C/593°C to Matsuura Power Station No.2 Unit, which will result in the significant improvement of plant efficiency.

Refer to Fig.1 for the location of the Matsuura Power Station.

The erection schedule of this No.2 Unit is briefed in Fig. 2 and in September 1996, main erection work was nearly complete. The boiler was firstly fired at the end of the year and will commence commercial operation in July, 1997.

This paper presents the main design features of Matsuura No.2 boiler, highlighting the application of an advanced steam temperature of 593°C/593°C at the turbine inlet.

2. IMPROVEMENT OF STEAM CONDITION

Fig. 3 describes the general trend of utility power plants in Japan. After the oil crisis in 1970's, Japanese utility companies tried to diversify fuel sources so as to achieve a stable supply of electricity, and imported coal is currently dominant in new installations. However, steam conditions had remained virtually the same for more than twenty years up until 1980's.

Higher steam conditions were initiated by global environmental issues, i.e. to reduce air pollutants especially CO_2 emissions by improving plant efficiency. Fig. 4 shows a record of steam parameter improvements established by BHK in Japan. The first 'USC' plant in Japan was built in 1989 employing gas fired boilers with steam conditions of 31MPa/566°C/566°C/566°C (2), (3). Then the newly installed coal fired plants had a typical live steam pressure of 24.1MPa, though steam temperature improved step by step. The most advanced steam condition currently in commercial operation is 24.1MPa/566°C/593°C, which was applied to Nanao-Ohta No.1 boiler of Hokuriku Electric Power Company supplied by BHK in 1995 (4).

This trend will continue with the plants currently under design/construction, and Matsuura No.2 Unit, the steam parameters of which are 24.1MPa/593°C/593°C, will achieve the highest live steam temperature in Japan in 1997.

Furthermore, subsequent units to be completed after 1997 will have slightly higher steam conditions as shown in Fig. 4. And the power plants of the next generation are expected to have more advanced steam conditions. Fig. 5 shows typical efficiency improvements by applying advanced steam conditions (5).

3. LEADING SPECIFICATIONS OF MATSUURA NO.2 BOILER

The plant is designed mainly for middle load operations, i.e. frequent and rapid load changes, quick start-up and shutdown are expected during the life of the plant. Sliding pressure operation will be carried out to improve plant efficiency at partial loads and to relieve turbine thermal stress.

Leading specifications of Matsuura No.2 boiler are summarized in Table 1 in comparison to No.1 boiler and design coal data of this boiler is listed in Table 2. Fifteen bituminous coals are specified for design coals and various coals from Australia, South Africa, China, the USA and Indonesia of wide range of characteristics will be burned by this boiler.

4. DESIGN FEATURES OF BOILER

A Side view of Matsuura No.2 boiler is shown in Fig. 6. As a middle load plant which burns a wide range of coals, the following considerations are made in boiler design.

(1) Spirally wound waterwall is applied to achieve a uniform temperature distribution at the furnace outlet throughout operating load range.

(2) Multi-lead ribbed tubes are used for spirally wound waterwall to assure adequate heat transfer in the combustion zone where heat flux in the furnace will be at a maximum.

(3) A three stage spray attemperation system is applied to improve steam temperature controllability during frequent and rapid load changes, and also to absorb the difference of heat absorption patterns in the boiler due to coal characteristic differences and slagging/fouling on the heating surfaces.

(4) The parallel gas biasing system together with gas recirculation is used to improve reheat steam temperature control, and spray attemperation is located in the intermediate stage of reheater to obtain better responses.

(5) The convection pass wall is of water cooled up-flow circuit type and consideration is made to stabilize flow balance inside each tube in sub-critical pressure ranges at partial loads.

(6) The water circulation lines consisting of steam/water separators, a drain tank and a boiler circulation pump to keep the minimum flow for furnace protection at low loads below the minimum once-through load of 25%MCR. This system makes the boiler start-up and shutdown procedure very simple, and keeps heat loss at a minimum.

(7) The high pressure and low pressure turbine bypass system is provided to reduce start-up time.

5. MATERIAL OF PRESSURE PARTS

Research and development of high temperature material has been carried out by steel suppliers, boiler manufactures and universities through Japan. Development progress of ferritic and austenitic materials are shown in Fig. 7 (6) and Fig. 8 respectively.

For high temperature headers and pipings for superheaters and reheaters, P91 (9Cr1MoVNb) developed by Oak Ridge National Laboratories was selected considering its high strength and wide in field experiences. Since late 1980's this material has been widely used for newly installed high temperature headers in Japan, and for the replacement of superheater headers in Germany, the UK and the USA.

On the other hand, new austenitic material has been applied to high material temperature superheater tubes, such as SUPER 304H (18Cr9Ni3CuNbN) developed by Sumitomo Metal Industries Ltd. (7). As listed in Table 3, this 18-8 stainless steel does not contain very expensive alloying elements such as Mo, W and has been found to be economical. This steel has extremely high creep rupture strength and allowable stress at 600~700°C more than 20% compared to TP347H. Refer to Fig. 9 for comparison.

Table 1 Leading specifications of Matsuura No. 2 Boiler

Item	No. 2 Boiler	No. 1 Boiler
Plant Output	1000 MW	1000 MW
Boiler Type	Benson type boiler	UP type boiler
Operation	Sliding pressure operation	Constant pressure operation
Main steam flow	2950 t/h	3170 t/h
Superheater outlet pressure	25 MPa	25 MPa
Superheater outlet temperature	598 °C	543 °C
Reheater outlet temperature	596 °C	569 °C
Feedwater temperature	288 °C	288 °C
Waterwall arrangement	Spirally wound wall	Vertical wall (UP - UP circuit)
Superheater steam control	Water - fuel ratio	Water - fuel ratio
	Three stage spray attemperation	Single stage spray attemperation
Reheater steam control	Parallel gas biasing	Parallel gas biasing
	Gas recirculation, Intermediate spray attemperation	Gas recirculation
		Inlet spray attemperation
Combustion system	Opposed wall firing	Opposed wall firing
Pulveriser	MPS118×7 (1 spare)	MPS118×7 (1 spare)
Burner type	Hitachi NR2 burner×56	Hitachi NR burner×70
Boiler outlet NOx (@6%O2)	180 ppm	200 ppm
Unburned carbon in ash	4%	5%

Table 2 Design coal analysis data

		China	Australia						USA			South Africa			Indonesia	
		A	B	C	D	E	F	G	H	I	J	K	L	M	N	O
GCV	kcal/kg	6930	6530	6790	6520	6850	6580	6560	6920	6510	6680	6400	6390	6610	6630	6820
Total Moisture	W(%)	10.8	10.0	10.6	18.0	8.2	10.0	9.8	13.4	15.9	8.5	10.3	10.8	10.9	15.8	13.6
Inherent Moisture	W(%)	4.1	3.2	3.9	8.0	3.4	3.2	2.6	6.9	9.6	2.9	3.5	4.1	4.2	10.1	6.7
Ash	W(%)	8.6	17.3	12.2	9.5	12.5	16.0	17.1	5.8	3.5	14.7	15.3	14.3	13.3	3.8	8.3
Volatile Matter	W(%)	28.2	29.8	28.9	25.8	32.3	30.7	28.4	40.6	37.6	34.7	23.9	24.7	31.6	43.5	44.5
Fixed Carbon	W(%)	59.1	49.7	55.0	56.7	51.8	50.1	51.9	46.7	49.3	47.7	57.3	56.9	50.9	42.6	40.5
Fuel Ratio	-	2.1	1.7	1.9	2.2	1.6	1.6	1.8	1.2	1.3	1.4	2.4	2.3	1.6	1.0	0.9
C	W(%)	76.1	66.6	72.3	75.7	72.7	69.3	68.8	74.0	74.6	70.1	70.1	71.1	71.2	74.5	71.3
H	W(%)	4.3	4.3	4.6	4.1	4.6	4.3	3.8	5.5	5.2	4.7	3.6	3.7	4.7	5.0	5.6
N	W(%)	0.8	1.6	1.6	1.7	1.5	1.5	1.4	0.8	1.4	1.4	1.8	1.7	1.6	1.7	1.2
O	W(%)	9.2	9.0	8.3	7.8	7.8	7.9	8.0	12.9	14.5	7.7	8.2	8.4	8.0	13.7	12.0
S	W(%)	0.6	0.6	0.5	0.4	0.4	0.5	0.5	0.6	0.4	1.0	0.4	0.2	0.7	0.9	1.0
Ash	W(%)	9.0	17.9	12.7	10.3	13.0	16.5	17.5	6.2	3.9	15.1	15.9	14.9	13.8	4.2	8.9
HGI	-	51	48	53	65	49	56	46	51	61	43	50	50	46	48	40

Table 3 Chemical composition of high strength stainless steel

Material	Main Composition	C	Si	Mn	P	S	Ni	Cr	Cu	Nb	N	Ti
SUPER304H	18Cr9Ni3CuNbN	0.07 ~ 0.13	≤0.30	≤1.00	≤0.040	≤0.010	7.5 ~ 10.5	17.0 ~ 19.0	2.5 ~ 3.5	0.3 ~ 0.6	0.05 ~ 0.12	~
TEMPALOY A1	18Cr10NiTiNb	0.07 ~ 0.14	≤1.00	≤2.00	≤0.040	≤0.030	9.0 ~ 12.0	17.5 ~ 19.5	~	≤0.40	~	≤0.20

To apply SUPER304H steel the extensive study and testing were carried out by BHK to confirm its reliability (9). The creep rupture strengths of parent metal and welded joints were tested. Fig 10 shows that this stainless steel has very high creep rupture strengths and those of welded joints are the same as the average value of the parent metal. It was confirmed that there is no tendency for reduction of the rupture strength of welded joints.

As shown in Fig. 11, it was confirmed that high temperature corrosion potential is very low in case of coal fired boilers if sulfur content in gas is less than 1.2%. Fig. 12 shows that corrosion loss of SUPER304H increases sharply when SO_2 content in gas reaches 0.1~0.12% which is equivalent to approximately 1.2% sulfur in coal. The design coals of Matsuura No.2 boiler have low sulfur content as shown Table 2 due to strict requirements of sulfur oxides emissions and therefore, this material was considered suitable for use.

Inner oxidation scale thickness mainly depends on amount of Cr% in the steel and temperature. However, generation of the scale is kept minimum up to 700°C when stainless tubes are shotblasted as shown in Fig. 13.

Table 3 also shows chemical composition of TEMPALOY A1 (18Cr10NiTiNb) developed by Nippon Kokan K.K. (8), which has been used for high temperature reheater tubes. This austenitic material also has high strength under high temperature as shown in Fig. 9 and BHK had applied this steel to the advanced steam condition boilers before Matsuura No.2 boiler, whereby it was proved to be liable in giving satisfactory performance as well as factors costwise.

Fig. 14 highlighted the area where P91, SUPER304H and TEMPALOY A1 have been used at Matsuura No.2 boiler. Both stainless steels are shotblasted to restrain generation of steam oxidation scale.

6. COMBUSTION SYSTEM

Japan has the strictest regulations for the prevention of air pollution and to meet these requirements BHK has been making significant efforts in development of Hitachi-NR burner. The phenomenon of overshooting in the concentrations of combustion intermediates in a fuel rich flame which accelerates the decomposition of nitrogen oxides (NOx) efficiently, was introduced into a pulverized coal firing burner, namely 'In-Flame NOx reduction' Hitachi-NR burner, where the Flame Stabilizing ring is installed as a key device. Hitachi-NR2 burner is based on the enhanced In-Flame NOx reduction technology, which incorporates two newly developed innovative devices; i.e., PC Concentrator and Space Creator (10). Refer to Fig. 15 for its construction.

Fifty-six (56) Hitachi-NR2 burners are equipped at Matsuura No.2 boiler together with a two stage combustion system in a suitably demensioned furnace, so that it is possible to reduce NOx emissions to 180ppm (6%O2) at the boiler outlet and unburned carbon in fly ash down to less than 4% even if South African coal of low volatile and high nitrogen content is burned. Refer to Table 2 for the coal data.

To fully utilize the burner, it is essential to obtain very fine coal. Matsuura No.2 boiler has seven large capacity roller type pulverizers of MPS 118 type incorporating with rotating classifiers. Refer to Fig. 16.

7. ENVIRONMENTAL PROTECTION

Emission of pollutants allowed for the No.2 unit are tabled below.

SOx (Sulfur Oxides)	Less than 80 ppm (1 hour value on normal operation, dry base)
NOx (Nitrogen Oxides)	Less than 60 ppm (1 hour value on normal operation, 6%O2, dry base)
Dust	Less than 0.03 g/Nm^3 (6%O_2 dry base)

The following equipments are installed to achieve these emission levels.

(1) Electrostatic Precipitator

This boiler will burn many kinds of coals imported from all over the world and most of them are have a sulfur content. To reduce variation of fly ash electric resistance, a Hot Electrostatic Precipitator has been selected by EPDC who confirmed satisfactory results of this design at Matsushima No.1 & 2, Takehara No.3 and Matsuura No.1 Units.

(2) Flue Gas Denitrification Plant

Due to the use of a Hot Electrostatic Precipitator, it is made possible to apply a low dust loading type Selective Catalyst Reactor, which has been supplied by BHK. A dry ammonia selective contact deoxidization method was used and plate type catalyst was selected for this process which demonstrates low pressure drop and little or no abrasion nor plugging by dust in various plants. Amount of catalyst has been decided to be able to control ammonia slip to a very low level so that pressure drop increase which is due to the production of ammonium bi-sulfate (NH_4HSO_4) in the airheater elements can be avoided.

(3) Flue Gas Desulfurisation Plant

The wet limestone-gypsum method was applied to Flue Gas Desulfurisation plant of one tower design where dust and sulfur oxides are removed simultaneously. To meet the requirement of sulfur oxides emissions, sulfur contents of coal are limited to less than 1.2% when selected.

8. DIRECTION OF FUTURE PLANTS

As prevosusly discussed, steam conditions in the future plants will no doubt advance progressively, and BHK has been in the forefront of continuous research and development of high strength steels with EPDC and steel suppliers.

BHK is currently proceeding with a feasibility study for application of new ferritic material of HCM12A (11) and NF616 (12) to 1000MW class coal fired boilers, which are now ready to be used for high temperature headers and pipings. Furthermore, BHK is studying the application of newly developed high strength materials such as ferritic 12%Cr materials and austenitic steels of HR3C, NF709 and others, the selection of which will depend on steam conditions and fuel types.

BHK trusts that these efforts will make it possible to advance steam conditions in power plants to the levels such as 30MPa/630°C/630°C in the near future.

9. CONCLUSION

Advanced steam conditions have played a key role in meeting increased electricity demands while reducing pollutant emissions and keeping up with global trend that demands improvements in efficiency of power plants.

Matsuura No.2 boiler for Electric Power Development Co., Ltd. which in the supercritical sliding pressure operation boiler designed for middle load usage, has advanced steam conditions of 24.1MPa/593°C/593°C to improve plant efficiency. High strength steel of austenitic SUPER304H for superheaters and TEMPALOY Al for reheaters adequately utilized to maintain the reliability of the plant.

In addition special attention was paid to protect the environment and the state-of-the-art technologies such as Hitachi-NR2 burners and flue gas denitrification plants were applied to keep pollutant emissions level at a minimum. Babcock-Hitachi K.K. will continue to play a key role in the development of boilers for advanced steam conditions in the future.

REFERENCES

(1) M. Uchiyama, Y. Koda, Electric Power Development Co., Ltd. Matsuura Thermal Power Station No.1 Unit Trial Operation Results of 1000MW with Imported Coal Firing Plant, The Thermal and Nuclear Power, 1991, Vol.42, No.4, Page 18–32.

(2) K. Nakamura, Construction Plan of Ultra Super Critical Pressure Plant – Kawagoe Thermal Power Station with Generating Capacity 700MW, The Thermal and Nuclear Power, 1987, Vol.38, No.8, Page 13–21.

(3) K. Iwanaka, Construction and Trial Operation of Kawagoe Thermal Power Station Unit No.1 and No.2, The Thermal and Nuclear Power, 1990, Vol.41, No.4, Page 45–57.

(4) T. Yasui, Characteristics and Operating Results of Nanao – Ohta Thermal Power Station Unit No.1, The Thermal and Nuclear Power, 1996, Vol.47, No.4, Page 35–43.

(5) K. Muramatsu, Development of Ultra Super Critical Plant, JHPI, 1996 Vol.34, No.2.

(6) Bendick, Vaillant, Rosselet, Hold, Blum, Newly Developed High Temperature Ferritic-Martensitic Steels from USA, Japan and Europe, VGB Conference, Kolding, 1993.

(7) Sawaragi, Ogawa, Kato, Natori, Hirano, Development of the Economical 18-8 Stainless Steel for Fossil Fired Boilers, The Sumitomo Search, 1992, No.48, Page 50–58.

(8) Minami, Kimura, Tanimura, Creep Rupture Properties of 18 Pct Cr8 Pct Ni-Ti Nb and Type 347H Austenitic Stainless Steels, J. Materials for Energy Systems, 1985, Vol.7, No.1, Page 45–54.

(9) Tamura, Sato, Fukuda, Mitsuhata, Mimura, Yamanouchi, Application of High Strength Heat Resistant Materials for Pressure Boundaries of Ultra Super Critical Boilers, Power•Energy Technology Symposium, Kobe, 1994.

(10) Koda, Morita, Kiyama, Yano, Baba, Kobayashi, Update '93 on Design and Application of Low NOx Combustion Technologies for Coal Fired Utility Boilers, NOx Symposium, Florida, 1993.

(11) Iseda, Natori, Sawaragi, Ogawa, Masuyama, Yokoyama, Development of High Strength and High Corrosion Resistance 12%Cr Steel tubes and Pipe (HCM12A) for Boilers, The Thermal and Nuclear Power, 1994, Vol.45, No.8, Page 54–63.

(12) T. Fujita, Current Progress in Advance. High Cr Ferritic Steels for High-Temperature Applications, ISIJ International, 1992, Vol.32, No.2, Page 175–181.

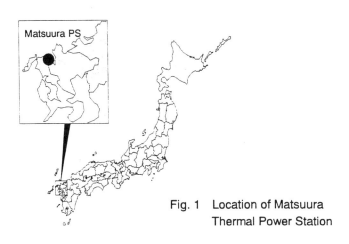

Fig. 1 Location of Matsuura Thermal Power Station

Year	1993	1994	1995	1996		1997
Month	NOV.	JUN.	JUL.	APR.	NOV.	JUN.
Milestone	C/V	F/C	H/L	H/T	I/F	C/O
Schedule	▼	▼	▼	▼	▼	▽

C/V : Civil work of main building F/C : First column erection
H/L : First Header lifting H/T : Hydrostatic Test
I/F : Initial Firing (First ignition) C/O : Commercial Operation

Fig. 2 Erection schedule of Matsuura No.2 Unit

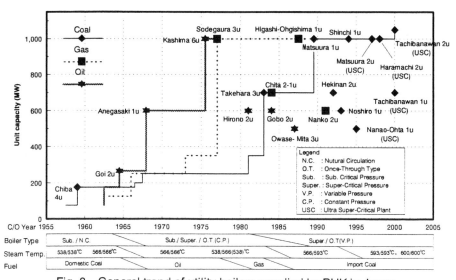

Fig. 3 General trend of utility boilers supplied by BHK in Japan

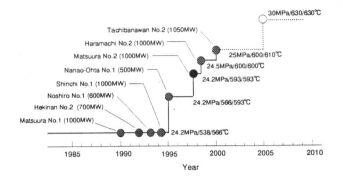

Fig. 4 Improvement of steam conditions in Japan

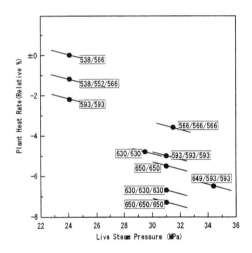

Fig. 5 Efficiency improvement by applying advanced steam condition

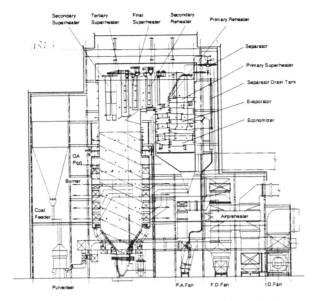

Fig 6 Boiler side view of Matsuura No. 2 Unit

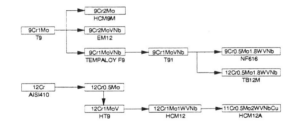

Fig. 7 Development progress of 9~12% Cr ferritic materials

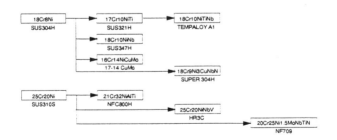

Fig. 8 Development progress of 18~25% Cr austenitic materials

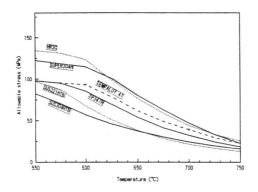

Fig. 9 Comparison of allowable stress of stainless steels

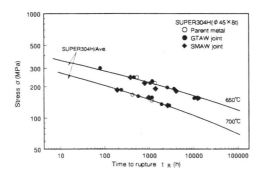

Fig. 10 Creep rupture strength of SUPER304H

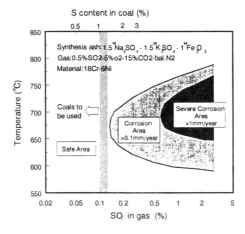

Fig. 11 High temperature corrosion characteristics of coal ash for stainless steel

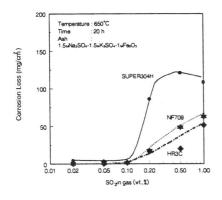

Fig. 12 Influence of SO$_2$ content on corrosion loss

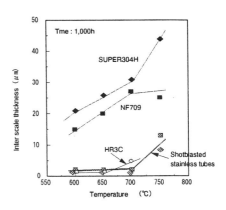

Fig. 13 Influence of steam temeprature on oxidation scale generation

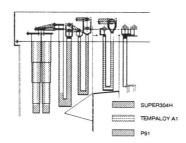

Fig. 14 Area of high strength material at Matsuura No. 2 boiler

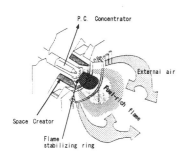

Fig. 15 Hitachi-NR2 burner

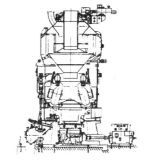

Fig. 16 MPS118 roller type pulverizer

C522/025/97

Two pass boiler design for advanced steam conditions

I R TORKINGTON CEng, **M UPTON,** and **F GEORGE** BSc
Mitsui Babcock Energy Limited, Crawley, UK

Synopsis

The future for fossil fuel fired utility boilers demands high efficiency plant operating on advanced steam cycles. This paper identifies the current limiting factors in boiler design and explores how plant design and boiler arrangement can be used to further extend the practicality of advanced steam cycles.

The two pass boiler will form the focus for this paper, however, alternative boiler arrangements will be compared and solutions for achieving overall cycle efficiency improvements will be considered within each of the boiler arrangements and advantages discussed.

Finally ideas will be proposed how new utility boiler designs should be developed to meet the future demands of advanced steam conditions.

1.0 Introduction

One of the biggest issues facing the purchasers of generating plant today is how to retain the well proven and familiar operating characteristics of pulverised coal fired steam cycle plant and at the same time meet increasingly demanding limits on polluting emissions. There is much less operating experience with other methods of coal combustion which offer inherently lower polluting emissions such as fluidised beds and gasification processes, and the steam cycles in which they need to be arranged. This is especially the case with large scale plant which is favoured in many parts of the world for electricity generation to achieve lower electrical energy unit costs by reduced manning levels and installation cost.

Despite these perceptions, the manufacturers of steam plant recognise clearly that they can only continue to meet the requirements and conservatisms of their customers, the electricity generators, for pulverised coal fired steam plant, if its performance can match that of alternatives. This has lead directly the search for pulverised coal fired steam cycles capable of delivering higher thermal efficiencies and lower through life fuel costs to off-set the costs of meeting emission limits.

This paper looks at the development of steam generator design to deliver steam at higher pressures and temperatures and the integration of such designs with the other steam cycle components so as to meet emission limits in the context of advanced steam cycles.

2.0 Raising Efficiency

The future role of the boiler designer is to both reduce the losses associated with the generation of steam from fossil fuels whilst offering arrangements that permit the use of advanced steam conditions. The performance of steam turbines is beyond the scope of this paper.

2.1 Boiler Thermal Efficiency

Improvements to the boiler thermal efficiency by reducing final gas temperature, lower unburned losses, lower excess air, improved leakage sealing in regenerative airheaters and reduction of boiler auxiliary power consumption all have a role to play in the development of boilers for the future.

How boiler thermal efficiency varies with excess air and exit gas temperature is shown in Figure 1 and it can be seen that the lower the excess air and lower the exit gas temperature then the better the boiler efficiency.

The pursuit of these gains is limited by combustion requirements where the need to ensure adequate combustion for good burn out and reduced unburned losses demands a minimum level of excess air in the furnace. The selection of exit gas temperature will be discussed in the next section. Leakage in the airheater will have a similar effect on efficiency as excess air for a given gas temperature and practical solutions to reduce leakage continue to be developed.

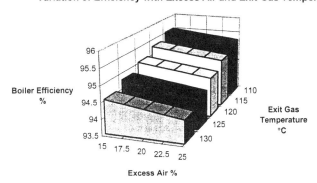

Figure 1

The electrical power consumption of the boiler auxiliaries such as pumps, fans and mills is a major consideration. Some of these auxiliaries, such as forced draught fans and mills, return a portion of the absorbed power to the steam cycle and their effect may be considered less important than induced draught fans. However, the compression heating in fans delivering high pressure primary air prior to air heating is unfavourable for airheater design and performance.

2.2 Steam Cycle Efficiency

Whilst reduction of losses offers improvements to the boiler efficiency, it is the advantages to be gained from advanced steam conditions, in increasing the available energy within the steam cycle that offers the greatest potential.

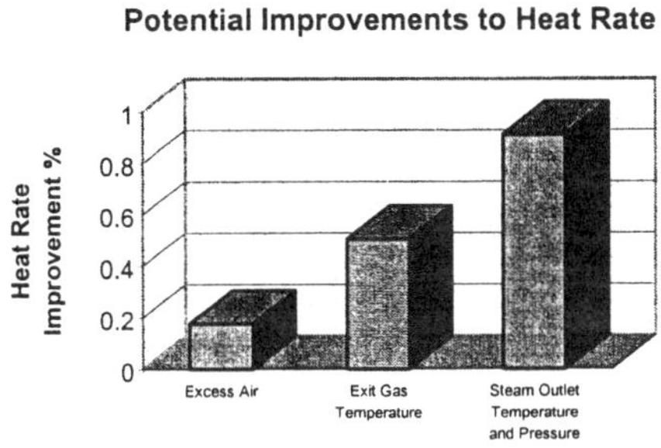

Potential Improvements to Heat Rate

(bar chart: Heat Rate Improvement % vs Excess Air, Exit Gas Temperature, Steam Outlet Temperature and Pressure)

Figure 2

Figure 2 compares the improvements to heat rate that can be achieved by reducing excess air from 25% to 18% and exit gas temperature from 120°C to 110°C with the potential advantage of taking steam conditions from 540°C at 250 bar to 600°C at 300 bar. The overall drive is therefore, towards the advanced steam conditions.

It is from this background that the boiler design engineer can play a role in providing plant that can deliver steam at greater temperatures and pressures with safe, reliable, flexible, efficient and environmentally sympathetic plant.

3.0 Factors Influencing Selection of Final Gas Temperature

We have seen how lower exit gas temperatures can be used to improve boiler efficiency and therefore station heat rate but there are several factors which influence this selection.

3.1 Feedwater Temperature

The first parameter that has a significant affect on the boiler design, which is often determined without reference to its impact on the design, is the feedwater temperature. This is usually established by the optimisation of the turbine steam flows and the use of bled steam within the feed heaters. It affects both two pass and tower boilers equally.

In current cycles the feedwater temperature is the lowest temperature fluid in the boiler and will therefore, set a practical limit to the gas temperature leaving the boiler. For economic use of heating surface a temperature differential between the incoming feed water and the leaving gas temperature must be maintained.

With the current supercritical steam cycles the feedwater temperature has risen above 300°C which affects the design of the Economiser, SCR (selective catalytic reduction) plant location, airheater design and furnace design. Furnace design will be covered in section 4.

3.2 Selective Catalytic Reduction

Current SCR plants reduce the concentration of NOx in the flue gas by the reaction of ammonia with a catalyst carried on a suitable substrate. This reaction operates at its optimum efficiency within a comparatively narrow temperature range (Figure 3). Lower efficiencies will require comparatively more ammonia being used to achieve the same reduction. This results in a higher proportion of ammonia being carried into the downstream plant, a phenomenon known as ammonia slip, which can lead to the formation and build-up of ammonium bisulphate on airheater elements, increasing gas and air pressure losses and adversely affecting thermal performance.

The natural characteristic gas temperature variation with boiler load means that at full load conditions gas passed to the reactor should be at the upper temperature limit whilst ensuring the lower limit is maintained at minimum boiler loads.

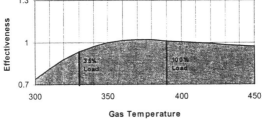

Figure 3

This window of operation identifies where in the gas stream the reactor should ideally be located if alternative methods of gas temperature control are not to be used. This places the reactor either upstream of the Economiser or located within the Economiser surface with some surface before the reactor and some downstream. The location is important to be able to choose the gas temperature entering the airheater unrestrained by the SCR reactor.

3.3 Economisers

The Economiser performs two functions: 1) it reduces the gas temperature to the required level for the airheater performance and 2) it acts as a buffer between the feed system and the furnace circuits. Some designers favour the removal of the economiser altogether. This may suit a tower boiler where the physical location of the Economiser is at the very top of the heating surface and therefore requires external pipework both to and from the inlet and outlet headers (Figure 4).

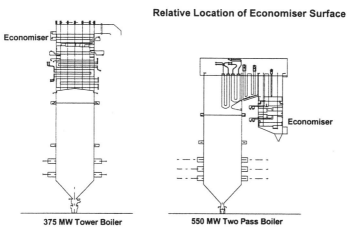

Figure 4

Removal, however, does mean that the furnace tubes are not protected from sudden thermal gradients due to feed heater trips. Economiser surface provides a buffer which suppresses temperature swings in the feedwater system before the water enters the furnace wall inlet headers and membraned furnace circuits. This reduces the potential for fatigue damage caused by thermal stress variations.

Economiser surface for economic design should maintain a gas temperature differential above the feedwater temperature of at least 30K. However, it and the other boiler surfaces must be selected to ensure that the water passed to the furnace circuit has a degree of sub-cooling.

3.4 Airheaters

As established above, the lower limit of gas temperature entering the airheater is controlled by the feedwater temperature. Acceptable outlet gas temperature is generally a function of the fuel constituents and acid dew point depending on the ash constituents and 110°C is now typical. The drive however is towards lower temperatures to improve boiler thermal efficiency, but advanced steam conditions are tending to take the airheater design to the practical limits of its own air side heat transfer efficiency.

The role of the regenerative airheater is to reduce the temperature of the gas from approximately 30°C above the feedwater temperature to the target for boiler efficiency. This is achieved by passing the combustion air through the heater matrix. However, the amount of air and the temperature at which it enters the airheater are constrained by other factors which affects the ability of the airheater to perform its function.

The amount of air is limited by the required combustion air, excess air and the proportion of bypass air needed to control the fuel pulverisers inlet and therefore product temperature to safe limits. This mill inlet temperature is a function of the fuel moisture, ambient air temperature and the type of mill. Figure 5 gives the mill inlet temperature required for various product temperatures and fuel moisture with an ambient temperature of 27°C, this is based on a vertical spindle mill with a fuel:air ratio of 1:1.7, variation of this ratio does allow some flexibility in the required mill inlet temperature but is constrained by transportation of the product and safety of the mill. Horizontal spindle mills which can operate at lower air:coal ratios, have not been considered due to their very high power consumption.

The permissible mill product temperature is dependent on fuel constituents particularly volatile content and mill manufacturers, especially of 'Ring and Roller' mills, need to be more flexible and realistic in their approach to product temperatures and fuel/air ratios. For bituminous coals the permitted product temperature can be in the order of 95°C to 100°C but will be limited to nearer 70°C for high volatile content coals such as those from Indonesia.

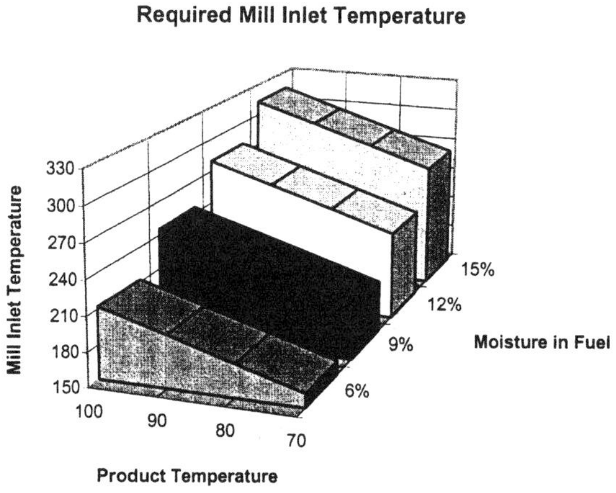

Required Mill Inlet Temperature

Figure 5

Ambient air temperature is fixed by the location of the plant and allowance must be made for temperature rise through upstream fans. The ability of the airheater to deliver air at the required mill inlet temperature can be limited by the airheater air side efficiency.

This is defined by the following equation:

Airheater air side efficiency = $\dfrac{t_2 - t_1}{T_1 - t_1}$

Where:
t_1 = air inlet temperature
t_2 = air outlet temperature
T_1 = gas inlet temperature

This has to be limited to about 93% otherwise the incremental change to the heat exchange surface required to achieve an improvement in performance becomes marginal and the attainment of that extra performance becomes uncertain.

If we consider a plant designed to achieve an airheater outlet gas temperature of 115°C and having advanced conditions, including a high feedwater temperature, the airheater will need to cool the flue gas leaving the economiser from 340°C, a drop of 225K. This will be achieved by heat transfer to the incoming combustion air. If it were possible to use all of the combustion air, the air side efficiency would be at a comfortable value. However, in this case, the resultant air temperature leaving the airheater, part of which is to be used for coal drying and transport in the milling system, exceeds the limit for safe mill operation.

In order to meet this limit a stream of air at lower temperature is needed. This is frequently achieved by mixing cold air from before the airheater, a process known as tempering. Whatever technique is used, the effect is to demand higher air side efficiency from the airheater. At the limiting value of 93%, the target design temperature of 115°C can no longer be met. The limit for the mill operation and therefore the airheater air side efficiency are dependant on the fuel moisture and permissible mill product temperature.

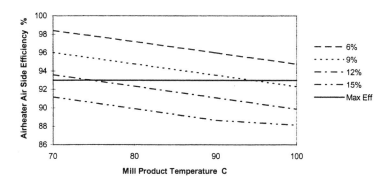

Figure 6

Figure 6 is based on the 340°C gas inlet and 115°C gas outlet case and it can be seen that fuel with only 6% moisture could not be used. To make it possible to use this fuel either the gas inlet temperature should be reduced, this has been set by the feedwater temperature, the ambient air temperature reduced, which is a function of geographical location, or the outlet gas temperature must be higher than 115°C which has an adverse effect on the boiler efficiency. 9% moisture coal can only be used with product temperatures above 95°C whilst 15% moisture coals require a greater temperature to achieve 100°C product temperature than the incoming gas for the case modelled.

From the above it can be seen that the heat and mass balance around the airheater and mills is a crucial part of establishing the practicality of advanced steam cycles.

From the steam cycle efficiency point of view it may be advantageous to drive the feedwater temperature higher and from a boiler perspective the reduction of final gas temperature with the use of different flue materials or linings may give better efficiency but the two cannot be considered in isolation and a total process approach needs to be taken.

In addition, if the power plant is to have flue gas desulphurisation (FGD) plant installed then consideration must be given to the final boiler gas temperature and its influence on stack temperature and plume formation. Gas/gas heaters may be used to exchange heat from the incoming gas to the FGD scrubber to the treated gas but again heating surface efficiency is involved and a minimum stack temperature must be maintained. The lower final gas temperatures favoured by the boiler designers for efficiency influence the downstream plant.

One option, which was considered for the Avedøreværket Blok II proposal for SK Energi in Denmark, was to use hot air from downstream of the airheater to boost the gas temperature just before the stack. This has the advantage of increasing air flow through the airheater thus enabling better optimisation of the airheater performance. Also the air boosting dilutes the gas thus reducing concentrations of emissions and gives freedom to operate without plumes which may be a consideration of any Environmental Impact Assessment (EIA). The negative side to this approach is the auxiliary power wasted by taking air at higher pressure than necessary. Although the final gas temperature can be lowered, the efficiency of the overall plant is not improved.

4.0 Furnace Design

4.1 Furnace Layout for Low NOx

Even if an SCR plant has been installed, it is far more cost effective to limit the production of NOx within the furnace than to remove it from the gas after formation as the consumables in the SCR process are expensive. For effective in-furnace NOx reduction the temperature within the combustion zone must be as low as possible but this affects the burn-out where higher temperatures promote good combustion. To ensure adequate residence time for burnout and low heat flux for low NOx, large furnaces with expanses of water cooled walls have become necessary.

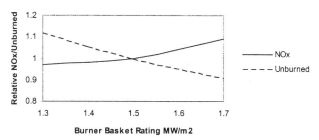

Figure 7

Figure 7 shows how burn-out and NOx production vary with 'burner basket zone' rating for a given furnace arrangement. Whilst this might appear to favour the once through boiler as furnace surface forms part of the superheater circuit, it does increase the heat absorbed by the furnace walls and raises their operating temperature.

4.2 Thermal Performance of Furnace

For once through boilers the total flow through the furnace tubes is less than the total evaporation of the boiler by the amount of the superheater temperature control spray flow. To achieve a suitable mass flux of water/steam in the furnace tubes in order to limit the tube temperature escalation due to deterioration of heat transfer coefficient as the boiling changes from nucleate to film, the number of tubes that make up the furnace walls is less than the number of vertical tubes which form the furnace walls in conventional drum boilers. The current solution to this is to incline the tubes and wind them in a shallow helix to form the walls. Clearly each tube is considerably longer than if it had taken a direct, vertical route as well as being of smaller diameter than the tubes in a drum boiler. In consequence pressure losses are much higher than they are in a drum boiler.

There have been vertical tube solutions to maintain high mass flows using multiple passes, but these have to operate at supercritical pressures at all times to avoid water and steam phases separating unequally at the circuit inlets and causing some tubes to be insufficiently cooled.

Each tube in a well designed helical winding, as well as being able to operate over a full range of pressures, embraces all the different zones of heat flux in the furnace. This is important in order to limit metal temperatures in the winding as the dynamic losses in the tubes lead to a negative flow response characteristic. That is to say that upsets in heat absorption lead to inverse changes in water flow and an exaggerated effect on exit temperature. This may be contrasted with a natural circulation furnace, in which flow responds in kind to higher heat absorption. Thus in a helical winding, full account of fluid temperature differences under off-design conditions must be allowed in selecting the tubing for furnace wall construction.

4.3 Furnace Wall Tube Layout

The helical winding continues to a level towards the top of the furnace where a transition occurs to vertical tubes, for the tower boiler enclosure or the two pass boiler open pass. For spiral designs control of the temperature unbalances requires methods of pressure balancing and positive flow mixing. For advanced steam conditions the temperature of the water entering the furnace may be higher, the furnace larger and the unbalance is therefore of even more concern.

The next major development is the use of internally ribbed vertical tubes in the furnace which are arranged to exhibit a positive flow characteristic with a greatly reduced pressure drop. This will be further discussed in section 4.6.

The wall temperature of the furnace tubes is a critical element in boiler designs for advanced steam conditions due to the materials of manufacture and any method of limiting this temperature or reducing the upset differentials is one key to opening up the possibilities of advances in steam conditions. By reducing inter tube temperature differentials it is possible to design the furnace tubes with metal temperatures closer to the bulk fluid temperature. The inlet conditions to the furnace walls are dictated by the feedwater temperatures and, the need for some economiser surface as described earlier, increases the temperature of the water entering the furnace circuits.

4.4 Furnace Tube Materials

Furnace walls in general use materials for which pre and post weld heating is not required. The best grade currently available for filling these requirements is 1% Cr ½% Mo, grade 620 British Standard steel (13 Cr Mo 44 DIN material). Higher grade materials can be formed into the necessary sections but welding of panels together at site to form the furnace enclosure which requires heat treatment of the entire panel is not currently economically feasible.

Whilst study and research continues into extending ferritic steels without the need for heat treatment it is the role of the boiler designer to find new ways to extend the capability of boilers using existing materials. There is a Japanese material HCM 2a under development which is a $2^{1}/_{4}$ % Cr ferritic steel that does not require heat treatment but this as yet does not have code approval.

4.5 Safeguarding the Furnace at Start-Up

Once through boilers, due to their limited number of tubes in the furnace, require a minimum flow through the furnace tubes at all times. The exact quantity of this minimum flow is dependant on the tubes, and the wall construction but is typically 35% to 40% of full load flow for spiral furnaces and is known as the Benson load. At loads below the Benson load, which is the load where the minimum flow equals the boiler evaporation, the water is circulated back to the economiser inlet. In order to achieve this circulation separator vessels are required which separate steam from the water, see Figure 8.

Typical Once Through Circuit With Recirculation

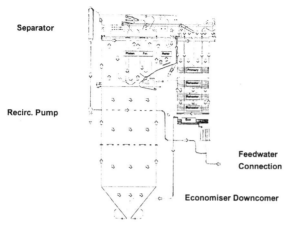

Figure 8

Separator vessels are generally located in the circuit at a point that gives some small degree of superheat at the Benson load (Figure 9). This position then gives the design temperature for the wall tubes when considered at the Boiler Maximum Continuous Rating load. The previously identified furnace wall materials can be used satisfactorily up to approximately 500°C mid wall metal temperature which is a function of the heat flux and the steam temperature allowing for temperature unbalances.

Pressure Enthalpy Diagram Showing Full Load and Benson Load

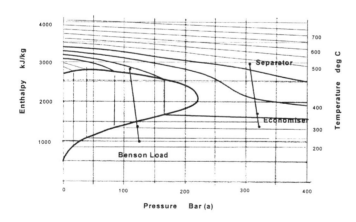

Figure 9

For tower boilers the optimum location in the circuit for the separators occurs at a position in the furnace below the horizontal surfaces and as the separator vessels require to be above the top bank of economiser for venting this necessitates multiple interconnecting pipework external to the boiler.

The steam from the separator is returned back to the same point in the circuit to continue in vertical enclosure wall tubes to the boiler roof which is uncooled. This arrangement has the advantage of being able to fully mix the steam circuits before entering the enclosure walls. The main disadvantage is that the separator's connections are positioned within the furnace and, during start-up, the spiral furnace has water circulating through it whilst the vertical enclosure tubes are uncooled.

In a two pass design, however, the spiral to vertical transition is arranged in the furnace at a position of higher gas temperature but the separators are located after the upper furnace open pass walls. This maintains cooled furnace walls and transitions throughout start-up and the spiral to vertical transition is used for steam/water mixing.

For either design the concern is to maintain an even controlled steam temperature at the point of highest metal temperature. Recently this has been assisted by the provision of a pressure balancing ring (Figure 10) connected to each of the spiral tubes at a level close to the top row of burners.

Pressure Balancing Ring

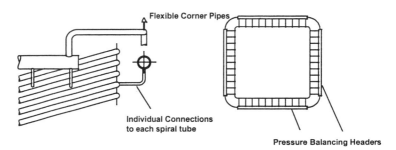

Figure 10

This has the effect of reducing unbalance caused by the negative flow characteristics of the spiral configuration and uneven lengths due to the route of tubes through burner openings. For a tower boiler the highest tube temperatures for the panel wall is at the separator connections at the spiral outlet in a zone of high heat flux whereas for the two pass design the critical tube temperatures occur at the top of the open pass, see Figure 8, where heat fluxes are lower and the mid wall tube temperatures are closer to the bulk steam temperatures.

Gas recycling can also be of benefit in controlling the furnace tube temperature unbalances as GR tends to inhibit heat absorption in the furnace.

4.6 Vertical Tube Furnaces

The introduction of internally ribbed tubes into the furnace allows the water/steam mass flux to be reduced whilst ensuring the required cooling of the tubes. The lower mass flux means more tubes can be placed in parallel for a given evaporation to the extent that it allows the furnace to be constructed of vertical tubes with sufficiently low pressure drop to give a positive flow characteristic.

The vertical tube furnace is a development that will enable once through supercritical boilers to operate with lower absorbed power due to reductions in furnace water/steam side pressure drops and will allow lower Benson loads to improve operational flexibility. The reduced pressure drop, by as much as 8 to 10 bar has a positive effect on the overall station heat rate. It will also remove the need for unheated strap support systems, as required by spiral designs, that limit the start-up times

For advanced steam cycles it is again of most use in combating the limiting factor of the furnace tube temperature. The positive flow characteristic, that is a feature of the vertical tube furnace design, automatically compensates for variations in furnace absorptions. Where the negative flow characteristic of the conventional spiral furnace requires pressure balancing and positive mixing methods.

This furnace tube arrangement favours neither the tower nor the two pass design but allows boiler manufacturers to build on their own experiences with natural circulation boilers in developing supporting and framing systems.

4.7 Furnace Exit Superheaters

The main advantage of the two pass boiler design is its ability to allow the inclusion of high temperature heating surface into areas of higher gas temperatures.

The initial ash deformation temperature (IDT) of the fuel being fired is the measure by which the hot flue gas can be passed over heating surface which may begin to collect slag deposit. At gas temperatures above the IDT of the fuel, ash is molten and can stick to surfaces it encounters. If the arrangement of surface is such that crevices or horizontal ledges are present then the deposit can build very quickly and will not be easily dislodged by conventional sootblowing.

For the tower boiler the first surface encountered is the screen of sling tubes turning towards the front and rear walls. Due to the circuitry of tower once through boilers these slings form part of the superheater circuit. The first major tube bank encountered is usually the secondary superheater, both this and the screen offer horizontal tubes which despite the use of wide cross pitches offers keys to promote slag build up. Current tower designs are exceeding IDT by in excess of 100K, Figure 11, in order to ensure adequate temperature head to meet the steam conditions required. The surface all requires supports within the gas stream and for the high temperature zones these are manufactured from high quality materials.

Heating Surface Arranged at Furnace Exit

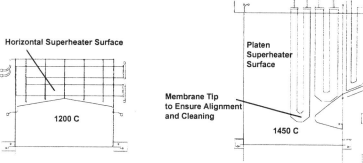

Figure 11

For the two pass design the inclusion of a Platen Superheater allows the secondary superheater to be located in gas temperatures up to 1500°C and with its vertical legs, membraned tip to ensure tube alignment, figure 11, and wide cross pitching can very readily be cleaned by conventional sootblowing. The furnace exit gas temperature can be controlled to below the IDT of the fuel for this arrangement and the use of pendant final superheater and final reheater means all supports for heating surface within high gas temperature zones are outside the gas stream.

Extensive experience has been gained with the use of platen superheaters in the open pass of both once through and natural circulation boilers, see Figure 12.

Platen Superheaters In Boilers Designed by Mitsui Babcock Energy

Power Station	No and Size	Platen Inlet Temp °C	Final Inlet Temp °C	Coal Ash IDT °C	Boiler Start-Up Date(s)
Willington	2 - 200	1350	1152	1150 - 1100	1960 - 1961
West Thurrock	3 - 300	1435	1435	1150 - 1100	1964 - 1965
Thorpe Marsh	1 - 550	1454	1149	1200 - 1050	1965
Ferrybridge	4 - 500	1474	1079	1200 - 1020	1966 - 1967
Kagisza	2 - 120	1322	1054	980	1966 - 1967
Ratcliffe	4 - 500	1474	1082	1200 - 1020	1967 - 1969
Esbjerg Blok 2	1 - 250	1454	1143	1082	1969
Fynsvaerket	1 - 220	1393	1116	1100	1968
Stalowa Wola	2 - 120	1332	1054	980	1968 - 1969
Didcot	4 - 500	1466	1071	1200 - 1020	1970 - 1972
Sierza	2 - 120	1332	1054	980	1971 - 1972
Drax	6 - 660	1477	1107	1200 - 1020	1972 - 1986
Tahkoluoto	1 - 220	1426	1152	900	1976
Matla	6 - 600	1473	1143	1170	1978 - 1983
Enstedvaerket	1 - 630	1509	1160	1200 - 1180	1979
Nijmegen	1 - 580	1500	1128	1075	1981
Castle Peak A	4 - 350	1483	1152	1350 - 1230	1982 - 1985
Castle Peak B	4 - 680	1480	1147	1200 - 1050	1985 - 1989
Lethabo	6 - 600	1398	1099	1190	1987 - 1992
Hwange	2 - 200	1490	1159	1380 - 1250	1987
Yue Yang	2 - 362	1518	1162	1500 - 1400	1991
Hemweg	1 - 650	1414	1136	1200 - 1080	1993
Meri Pori	1 - 600	1329	1070	1100	1993

Figure 12

These references include supercritical once through boilers in Finland and Holland which gives the necessary confidence to apply them to boilers for even more advanced steam conditions where they are essential to meet the required temperatures.

With the secondary superheater in the furnace the options occur where to locate the final superheater and reheater surface to maintain optimum temperature heads. In the two pass design the final surfaces can be placed in sequence and the gas side mass flux varied dependant on the velocity of the gas and erosive index of the ash by varying the furnace roof height which also controls heat pick up and therefore metal temperatures.

One feature of the tower boiler is its ability to have the final banks of both the superheater and the reheater in the same gas temperature zone by placing the surface back to back, see Figure 13, taking half the boiler depth each. This has the advantage of limiting the temperature pick up and therefore steam unbalances which help to control the ultimate metal temperatures. The only disadvantage is the potential for gas side unbalance that may be caused.

Arrangement of Superheater and Reheater Surface in a Tower Boiler

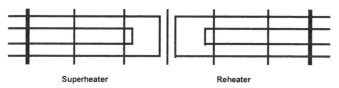

Superheater   Reheater

Figure 13

Interleaving banks is possible across the width of the boiler but with mass flux considerations and limiting of temperature pick-up and imbalance, the back to back arrangement offers greater advantages.

All the above assumes that materials are available to the boiler designer that allow steam at the final temperatures and pressures being considered. It is the subject of much research and development to find materials that offer the creep/fatigue capabilities required at these temperatures and pressures whilst being practical in their application. From the utility boiler point of view there is a practical limit to the ultimate steam temperatures that can be achieved when firing coal. The IDT is one factor that will limit the gas temperature at which practical heating surface can be installed, however, as can be seen from the above arguments, careful design of the furnace can extend the practical use of advanced steam cycles even in conjunction with today's available materials.

5.0 Reheaters

Advanced steam cycles may or may not use two stages of reheater and whilst it is clear that overall cycle efficiency can be improved in such a way there are a number of economic and thermodynamic constraints.

The additional cost of the turbine steam jacket required for two stages of reheat may outweigh the through life savings that the increased efficiency gives. Thermodynamically the arrangement of heating surface requires careful consideration with respect to suitable temperature heads.

However, double reheat has been used with good results and is being used in the CONVOY plants in Denmark and has a role to play in the future if reduced emissions is ultimately the driving force. Double reheat systems can be optimised to produce the most efficient use of any of the temperature control methods.

All normal methods of temperature control can be applied to the double reheater cycle, one of which is discussed below. However, if the unit is to be operated as a base load station, which for high efficiency plant is probable, then there may not be a need to control the reheater outlet temperature at all. At full load adequate surface is provided for full reheat temperature and at part load the temperature is allowed to fall. Care must be taken in the design to control the maximum wetness of exhaust steam from the turbine.

One possible arrangement is for the first stage to be controlled by spray attemperation and the second by gas recycling. As previously mentioned the gas recycling has the added benefit of reducing furnace heat absorption and thus helps protect the furnace tube material. It does, however, increase the gas mass flux and therefore the gas side heat transfer coefficient for all the convective banks. For the first stage of reheater this would mean the extra surface provided for spray control would all perform better, requiring more attemperator spray. This surface would therefore benefit from being of the plate gilled, extended surface type. In this way the additional surface is provided cost effectively with reduced bank height and better sootblower coverage. This method of control is available to both Two Pass and Tower boilers and there is a gas temperature limit for the inclusion of extended surface dependant on the material used. The plate gilled surface improves the gas side heat transfer coefficient as installed and the increased gas flow required by the second stage of reheater has less effect on the first stage as the steam side coefficient starts to govern the overall heat transfer coefficient. Even for single reheat cycle plant extended surface tubing can be utilised.

An alternative method would be to arrange each of the reheater stages in separate streams of a divided back pass boiler with damper biasing to control the relative temperatures. This arrangement is only possible for Two Pass Boilers.

6.0 The Future

In the short term boiler designers require suitable materials for construction of the furnace panels that do not need pre or post weld heating. This would immediately increase the temperatures and pressures for new plant as it would allow the full use of the currently available austenitic materials for the high temperature components. In the longer term it is the high temperature materials that will require development. It is also true that in order to justify the cost of their development and testing and ultimate approval into recognised international design codes the benefits from these materials must be significant.

It is also clear that advanced steam cycles need to consider all aspects of the plant including the sensitivity of the boiler to changes within the turbine and feed heating system. The current catalytic reactors for NOx reduction are limiting in their operating regime and development of low temperature catalysts would give the system more freedom with regard to feedwater temperature and airheater duty. A combining of DeNOx surface and regenerative airheater elements would offer an effective combination whilst reducing the draught loss and therefore the auxiliary power and heat rate of the plant

The two pass boiler offers advantages in the pursuit of the advanced steam conditions not purely for its arrangement of surface but also due to its compact design. Materials developed for main steam pipes will inevitably be expensive and methods to reduce the length of runs between boiler and turbine are required.

7.0 Conclusion

The efficiency of electricity generating stations using fossil fuels depends not on the isolated performance of the individual components of a station such as steam turbine, boiler and flue gas clean-up, but on the performance of the whole. Advancing steam conditions does not improve boiler efficiency, indeed this paper has shown that raising feedwater temperature can lower boiler efficiency. A close and sympathetic integration of the different parts is essential if advances are to be realised when working close to process and material limits.

The design and layout of boilers for advanced conditions must respect the limits of materials and process parameters whether these are hard limits imposed either by thermodynamic principles of fluid flow and heat transfer and by measured properties of materials or by the softer limits indicated by parameters such as combustion and ash deposit formation and management largely based upon experience. The layout of both two pass and tower boilers can be made in such a way to meet hard limits. The comparison between them must focus on the soft limits and it is the responsibility of designers to take account of the requirements of the operators throughout the life of the plant.

The layout of the two pass boilers with pendant high temperature tube elements offers particular advantages for advanced plant. Pendant tube elements provide a low strength bond between coal ash accumulations and tubes as well as an inherently robust method of support since longitudinal stresses in a vertical tube are very much lower than in horizontal tubes. Load carrying attachments for pendant elements can be and therefore are located safely outside the gas pass. These features enable heating surface to be arranged at the exit of the two pass boiler furnace in a region of high gas temperature. For advanced plant where furnace tube material limits are approached this arrangement accelerates the decrease of outside heat flux and raises the allowable steam temperature in furnace tubes. Horizontal surfaces which are an intrinsic feature of tower boilers are not suited to such a layout and restrict tower boilers operating conditions.

The two pass boiler design for advanced steam conditions thus meets all the hard limits arising from thermodynamics, heat transfer and materials as well as the soft limits of combustion and ash deposit management which are central to high availability.

Materials for Boilers

C522/027/97

New steels for advanced coal fired plant up to 620°C

E METCALFE BSc, PhD, W T BAKKER, R BLUM, R P BYGATE, T B GIBBONS, J HALD, F MASUYAMA,
H NAOI, S PRICE, and Y SAWARAGI
For authors' affiliations, see end of paper

SYNOPSIS   There are strong environmental and economic incentives to increase the thermal efficiency of fossil fired power stations, and this has led to a steady increase in steam temperatures and pressures. In addition, the economics of power generation are driving down the price of plant so that advanced technology must be available at reduced installed cost. The key to securing these conditions is the development of high temperature materials, available at an acceptable price, particularly for thick section components in the boiler and turbine. These considerations led to the establishment of a four year EPRI project with partners from Japan, UK, USA and Denmark whose objective was to establish strong 9Cr and 12Cr steels as practical, validated materials for thick section boiler components such as headers and main steam lines. This successful project has developed three strong steels for thick section components for plant operating in the temperature range 565 - 620°C and two of them, P92 (NF616) and P122 (HCM12A), have received ASME Code approval.

This project has now entered its second phase with the fabrication of full sized headers which have been installed in the NJV power plant in Denmark. In addition, there is further work on the long term microstructural stability of the steels, and a full sized pressure vessel test under accelerated conditions is about to start.

1 INTRODUCTION

The trend for future coal fired power plant is to reduce the CO_2 emission by increasing the steam parameters while driving costs down. Power plants with an efficiency of about 50% with steam parameters of about 310 bar and 600 - 610°C are being considered in a number of countries. However, the lack of suitable ferritic steels for the construction of thick section boiler components and steam lines has until now been the main factors limiting the realisation of these prospects.

In order to address the materials requirements for thick section boiler components, and to develop steels with significantly higher creep rupture strengths than Grade 91, an international consortium of steelmakers, boilermakers and electricity producers was established six years ago under EPRI Project RP1403. The companies in the project are those shown against the authors of this paper. The consortium set itself challenging target criteria for the new steels it wished to develop and validate (Table 1). Three W-strengthened steels, NF616, HCM12A and TB12M, have been developed and validated for thick section components (1) and sufficient long term creep rupture data have been obtained for NF616 and HCM12A for these two materials to gain ASME Code approval (2). This paper presents data relevant to all three steels but concentrates on the properties of NF616 and HCM12A because of their ASME Code approval and thus the immediate opportunities for use.

TABLE 1   Target Criteria for the Steels

PROPERTY	TARGET
100 000h creep rupture strength at 600°C	>140 MPa
100 000h cross weld rupture strength at 600°C (or 30% strength loss allowed)	>100 MPa
20°C impact strength (unaged)	>40 J
0.2% proof stress at 20°C	>400 MPa
0.2% proof stress at 600°C	>250 MPa
Ultimate tensile strength at 20°C	>600 MPa
Ultimate tensile strength at 600°C	>350 MPa
Elongation	>20%
Reduction of Area	>70%
Reduction of Area after 10 000h at 600°C	>40%
Minimum average tempering temperature (for thick section components a variation of $\pm 20°C$ will be allowed around this average value)	>760°C

2   MECHANICAL AND CREEP RUPTURE PROPERTIES

Large diameter thick walled pipes of NF616, HCM12A and TB12M of dimensions 350mm outer diameter and 50mm wall thickness were fabricated by Nippon Steel Corporation, Sumitomo Metal Industries, and Forgemasters Steel and Engineering Ltd respectively. Table 2 shows the specification and analysis of typical melts of NF616, HCM12A and TB12M. With regard to heat treatment NF616 was normalised at 1065°C x 2h AC and tempered at 770°C x 2h AC; for HCM12A it was normalised at 1050°C x 1h AC and tempered at 770°C x 7h AC, and for TB12M it was normalised at 1070°C x 4h AC and tempered at 770°C x 8h AC.   A further heat treatment of 740°C x 4h AC to correspond to stress relieving during fabrication was applied prior to mechanical property and creep rupture property determination. Ageing at 720°C for 200h was applied to some specimens to simulate long service times, although it was later discovered that this treatment did not simulate the microstructural developments at temperatures of 650°C and below (see Section 4).

The short term mechanical property values for all three steels met the target criteria.   For example, Figure 1 shows results of the elevated temperature tensile tests for NF616 and HCM12A pipes. No difference is observed for different sampling directions. The properties all satisfied the target criteria. Figure 2 shows the Charpy impact test results. Tests were performed on samples taken from both the tangential and longitudinal directions and no significant differences could be seen.

TABLE 2   Representative chemical compositions (wt%) and specifications

Element	NF616		HCM12A		TB12M	
	Analysis	Specification	Analysis	Specification	Analysis	Specification
C	0.106	0.07-0.13	0.12	0.07-0.14	0.13	0.10-0.15
Si	0.04	<0.50	0.05	<0.50	0.04	<0.50
Mn	0.46	0.30-0.60	0.64	<0.70	0.55	0.40-0.60
P	0.008	<0.020	0.016	<0.020	0.005	<0.020
S	0.010	0.010	0.001	<0.010	0.004	<0.010
Cr	8.96	8.50-9.50	10.61	10.00-12.50	11.07	11.00-11.30
Mo	0.47	0.30-0.60	0.44	0.25-0.60	0.50	0.40-0.60
W	1.84	1.50-2.00	1.87	1.50-2.50	1.82	1.60-1.90
Ni	0.06	<0.40	0.32	<0.50	0.75	0.70-1.00
Cu	-	-	0.86	0.30-1.70	-	-
V	0.20	0.15-0.25	0.21	0.15-0.30	0.20	0.15-0.25
Nb	0.069	0.04-0.09	0.05	0.04-0.10	0.080	0.04-0.09
N	0.051	0.030-0.070	0.064	0.040-0.10	0.046	0.040-0.09
Al	0.007	<0.040	0.022	<0.040	0.003	<0.010
B	0.001	0.001-0.006	0.0022	<0.005	-	<0.005

The creep rupture properties of NF616 and HCM12A have been determined from the extensive data packs which were submitted to ASME for Code approval, and analysed by standard techniques used by the Materials Properties Council of USA. For NF616, 284 data points from ten heats were analysed, with numerous stress rupture times exceeding 10 000h and some test durations exceeding 30 000h. A similar data pack for HCM12A was provided with the total number of data points being 290. The usual way of comparing the strengths of high temperature steels is to estimate by established extrapolation techniques the creep rupture strength after 100 000h at 600°C. The strengths predicted by the ASME analysis are 132 MPa for NF616 and 129 MPa for HCM12A, which are an improvement of about 40% over P91. The value of this improvement is shown in Figure 3 which compares the allowable stresses for NF616 (P92) and HCM12A (P122) with those for P91 (mod. 9Cr-1Mo) and P22 (2.25Cr-1Mo) steel, and also shows the sectional geometry of a main steam pipe designed for a pressure of 24MPa and temperatures of 570°C and 600°C.

3 FABRICATION TRIALS

3.1 Pipe Bending

In order to simulate fabricability using these materials pipe bending trials were carried out using induction bending for NF616 and HCM12A and flame bending for TB12M. In the induction bending process, the pipe was bent forward while the pipe circumference was being heated on its outside surface using a high frequency induction heating coil, and the pipes were bent to an angle of 90° at a bending rate of 0.3 mm.s^{-1} at 1000°C. After bending the pipes were normalised and tempered. In conventional fire bending the pipe is heated to 1200°C and then pulled to the required radius. Bending started at a temperature of 1050°C and was completed at a temperature of 920°C. After bending the pipe was normalised and tempered.

Dimensional measurements of the pipes before and after bending showed that the wall thickness had decreased at the extrados but increased at the intrados; however these changes were very similar to those seen in conventional materials like P91 and P22, and no particular problems were found with respect to bending formability. For the flame bent TB12M, the maximum thinning seen was 6.6% and the maximum ovality was 4.4%.

3.2 Weldability

All three steels exhibit very good weldability. For example Figure 4 compares NF616 and HCM12A with conventional P91 and P22 steel during Y-groove weld cracking tests (3). The preheating temperature for complete prevention of weld cracking for P22 was 300°C, whereas the corresponding temperature for the P91, NF616 and HCM12A steels was 200°C, showing these to have superior weldability compared to the low alloy steels despite the high chromium levels. The major reason for this may derive from their relatively low carbon levels of about 0.1%.

Use of these high strength steels as headers would involve dissimilar metal welds with austenitic stainless steel tubes. It is also likely that the steels would have to be welded to thick section components in P91. Procedures have been developed for similar and dissimilar metal welds in these steels using processes ranging from tungsten inert gas (TIG, also known as gas tungsten arc or GTAW), manual metal arc (MMA or SMAW) and submerged arc (SAW). For the ferrtic - austenitic tube welds, either an austenitic (Type 309) or a nickel based filler (Inco 82 or Inco 182) was used. For the pipe sized ferritic - ferritic welds with P91, either NF616 or a Grade 91 consumable was used. Most of the initial development work was done using a Grade 91 consumable but matching fillers are now available for both NF616 and HCM12A. In general, preheats in excess of 150°C and interpass temperatures of 300-350°C have been used. Stress relieving is considered to be mandatory for these steels, even in tube sizes, because as-welded tube joints cracked at the fusion line when bent. The welds are stress relieved at 740-750°C. In bend tests, all stress relieved welds satisfied the ASME Section IX acceptance criterion of a 180° angle bend without cracking.

4 MICROSTRUCTURAL STABILITY

Confidence in the long term microstructural stability of these newly developed 9 - 12%Cr steels which contain W is important. Isothermal ageing tests for up to 10 000 h at 600°C and 650°C have been carried out and comparisons made with Grade 91 steel. In addition, detailed microstructural characterisation and thermodynamic computer modelling of the microstructure has been carried out by Hald (4). Microstructural investigations show that the most evident change in microstructure during ageing is the precipitation of intermetallic Laves Phase $(Fe,Cr)_2(Mo,W)$ at 600°C and 650°C in the W-alloyed steels. Laves phase also formed in P91 but only at 600°C. With the aid of thermodynamic computer calculation, a model has been developed which fully describes the precipitation of Laves Phase in NF616 and HCM12A as a function of time and temperature, Figure 5. It should be noted that accelerated ageing at 720°C does not allow Laves phase to form and therefore does not simulate the long term microstructural development at lower temperatures and longer times. Comparative creep tests on NF616 and HCM12A materials which precipitate Laves Phase during creep, and on materials heat treated to precipitate all Laves phase before creep testing indicate that Laves phase precipitation during the creep process is the primary strengthening effect of tungsten in these 9-12%Cr steels and that the solid solution strengthening effect of tungsten is small.

This conclusion concerning the strengthening effect of W proposed by Hald (4) has recently been supported (5) by a comparison of the creep rupture properties of P91, E911 (a recently developed 9Cr-1M0-1W-V-Nb steel in the European COST programme), and P92. Part of the basic strength of the 9Cr steels is provided by the solid solution hardening elements Cr, Mo and W. If the major contribution of Mo and W is solid solution strengthening, then the effectiveness should decrease with increasing test time as Mo and W are taken out of solution due to the formation of the stable carbide M_6C and Laves phase. However the differences in rupture strength between the steels increase rather than decrease with test duration, suggesting that the beneficial effects of Mo and W do not diminish.

The rupture strength of P92 after ageing for 10 000h at 650°C, equivalent to 200 000h at 600°C, was similar to P91 in the unaged condition. This indicates that the creep rupture strength of the three W-containing steels does not decrease catastrophically due to the precipitation of Laves Phase and the extrapolated strengths can be regarded with confidence.

5 ECONOMICS OF USAGE OF THE NEW STEELS

Given the higher strengths of the new steels the wall thicknesses of critical pressure parts can be reduced resulting in savings of weight of any fabricated structure. In addition, reduced section thicknesses will result in lower temperature gradients through the wall and consequently a lower thermal fatigue loading on the component which is an advantage during load following operations. A simple comparison has been carried out on the basis of a superheater header designed to be fabricated in Grade 22, Grade 91 and HCM12A. This was a header with an overall length of 20m, a nominal outer diameter of 400mm and a weight of approximately 9000 kg. The design included two tees and two end caps. The comparison of relative costs compared to Grade 91 was made for an estimated weight of pipe of unit length using cost data supplied by Sumitomo Metal Industries (6) which showed that the prices of Grade 22 and HCM12A were 55% and 118% respectively compared to that of Grade 91.

The data on the relative weight and cost of HCM12A and Grade 22 compared to Grade 91 are shown in Figures 5 and 6 respectively for a design pressure of 31 MPa. The benefits of using the stronger steel HCM12A (or NF616) are evident and are more significant for the higher temperatures. For a temperature of 625°C, the cost of using HCM12A would be approximately half that of Grade 91.

6 FULL SIZE COMPONENT FABRICATION AND PLANT TRIALS

A key part of the second phase of the project has been the fabrication of superheater headers for installation in one of the ELSAM power stations. A 400 MW coal fired power plant, NJV, with double reheat and steam parameters 290 bar/580°C/580°C/580°C, is under construction by ELSAM and will be commissioned in 1998. ELSAM have replaced 3 of the 28 HP superheater steam outlet collectors made of P91 with 2 collectors of NF616 and 1 collector of HCM12A, in order to run an in-plant test for a period of approximately 5 years. Each header is of approximate dimensions 1800 mm long, 160 mm diameter, 45 mm wall thickness. The design of the headers is in accordance with the ASME Code using design parameters 602°C and 313 bar. The test headers are fitted with P91 transition pieces at the outlet and with P91 stubs for connection to the superheater tubes. The headers were subjected to a rigorous inspection during manufacture and prior to installation.

In total four headers were manufactured:

Mitsubishi Heavy Industries	one HCM12A header, one NF616 header
Rolls Royce ICL	one NF616 header
ABB Combustion Engineering	one HCM12A header

In order to evaluate material in more demanding conditions more typical of the eventual usage of these steels the header produced by ABB Combustion Engineering will be tested at high temperature and pressure in a test cell operated by Mitsubishi Heavy Industries. These accelerated test conditions will provide important information on the performance of ferritic - ferritic welds (45mm wall thickness) and on tube sized ferritic - austenitic welds which will further increase confidence in the reliability of these high strength materials.

7 CONCLUSIONS

This international project team, operating within an EPRI project, has developed three strong ferritic steels; NF616 and HCM12A, which have sufficient long term creep rupture data to have already gained ASME Code approval, and TB12M. All three steels have been subjected to a rigorous validation programme which includes welding consumable development and fabrication trials. In addition, full sized superheater steam outlet steam collectors have been manufactured and installed in a 400 MW power station. The steels are now ready for use to manufacture the high temperature plant components which are necessary to support the ultrasupercritical plant option.

REFERENCES

(1) Metcalfe, E, editor of New Steels For Advanced Plant up to 620°C, *The EPRI/National Power Conference, London, May 1995. Published by PicA, Drayton, Oxon OX14 4HP.*

(2) Masuyama, F, ASME Code Approval for NF616 and HCM12A, *The EPRI/National Power Conference on New Steels for Advanced Plant up to 620°C, London, May 1995, 98 -113*

(3) Masuyama, F, and Yokoyama, T, NF616 Fabrication Trials in Comparison With HCM12A, *ibid, 30 - 44*

(4) Hald, J, Materials Comparisons Between NF616, HCM12A and TB12M - III: Microstructural stability and ageing, *ibid, 152 - 173*

(5) Ennis, P. J. and Wachter, O, A comparison of the Creep Rupture Properties of the 9%Cr Steels, P91, E911 and P92, *VGB Conference "Materials and Welding Technology in Power Plants" Cottbus, 8-9 October, 1996, 4.1 to 4.13*

(6) Yang, Z, Fong, M. A, and Gibbons, T.B, Steels For Thick Section Parts: Comparison of economics of usage in a typical design, *The EPRI/National Power Conference on New Steels for Advanced Plant up to 620°C, London May 1995, 174 - 183*

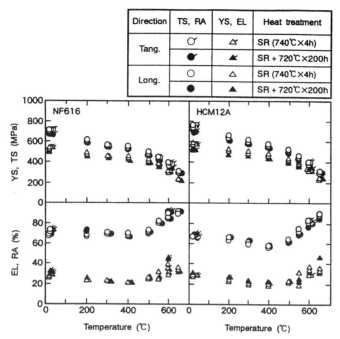

Figure 1 Tensile properties for NF616 and HCM12A

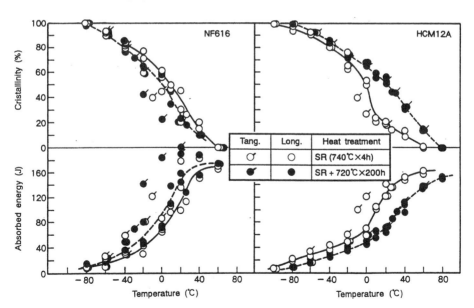

Figure 2 Charpy impact properties for NF616 and HCM12A

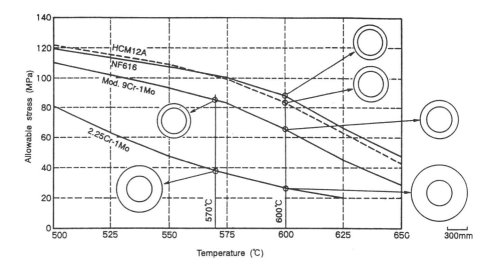

Figure 3 Comparison of allowable stresses and sectional view of main steam pipes designed at 570°C and 600°C

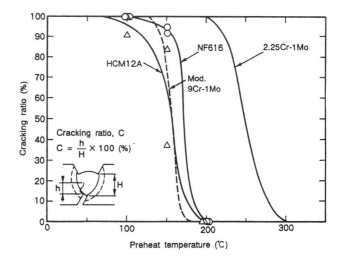

Figure 4 Y-groove weld cracking ratio versus preheat temperature relationship for NF616 and HCM12A compared to 2.25Cr-1Mo and modified 9Cr-1Mo (P91)

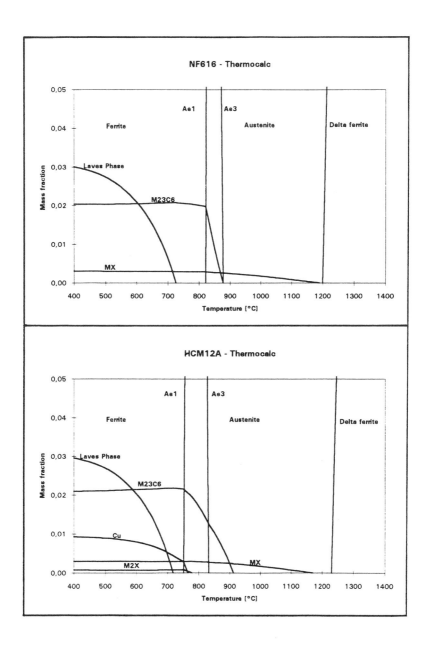

Figure 5   Equilibrium phases in NF616 and HCM12A calculated by Thermocalc

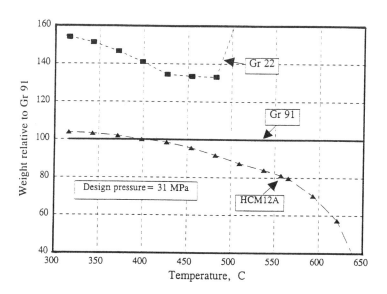

Figure 6   Relative weight compared to Grade 91 of headers manufactured in P22 (2.25Cr-1Mo) and P122 (HCM12A) for a design pressure of 31 MPa

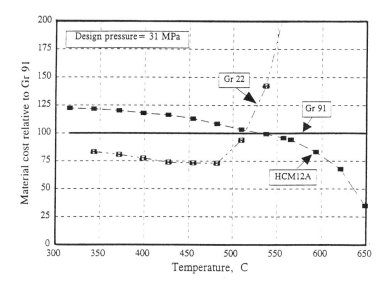

Figure 7   Relative cost compared to Grade 91 of a header manufactured in P22 (2.25Cr-1Mo) and P122 (HCM12A) for a design pressure of 31 MPa

Authors' Affiliations

E METCALFE
National Power, UK
W T BAKKER
Electric Power Research Institute, USA
R BLUM
ELSAM, Denmark
P BYGATE
Rolls Royce ICL, UK
T B GIBBONS
ABB Combustion Engineering, USA
J HALD
Technical University of Denmark, Denmark
F MASUYAMA
Mitsubishi Heavy Industries, Japan
H NAOI
Nippon Steel Corporation, Japan
S PRICE
Forgemasters Steel Engineering Limited, UK
Y SAWARAGI
Sumitomo Metal Industries, Japan

© with authors

C522/026/97

Crack stability assessment for advanced 9CR steels in boiler components

Eur Ing **N TAYLOR** BA, BAI, PhD, MIE(Ire), VDI, FEANI, **E LUCON**, and **V BICEGO**
CISE Spa, Milan, Italy
P BONTEMPI
ENEL Spa, Milan, Italy

ABSTRACT

The results of fracture toughness, fatigue crack growth and creep crack growth tests on specimens machined from modified 9Cr (P91) base material and welds as well as E911 (a 9%Cr alloy with added tungsten) are reported. No significant differences were found between the P91 parent and the fusion line specimens. E911's fracture properties are found to be at least as good as those of P91, underlining the promise of this alloy. Established creep and creep-fatigue crack growth analysis procedures have been applied to model frequency effects in FCG testing and time to rupture for CCG conditions. For the latter, consideration of primary creep behaviour is found to be important.

1. INTRODUCTION

Advanced high chromium ferritic alloys are finding increasing use in the drive to raise steam power plant efficiency. While investigations tend to focus on the tensile, creep and low cycle fatigue properties of parent material and welds, determination of fracture mechanics parameters also play an important part in establishing a complete suite of life assessment procedures. Defect stability analysis can serve for advanced design verification (e.g. assessing flaws introduced during production) or for dealing with maintenance programming/residual life issues should cracks be detected during operation. In this regard the use of sophisticated component monitoring systems (1) capable of detecting crack initiation events leads to the problem of evaluating the criticality of the situation with regard to safety and costs.

The overall requirements can essentially be reduced to 3 sets of parameters: fracture toughness as described by a K_{Ic} or J_{Ic} value, a fatigue crack growth law ($da/dN = S\Delta K^v$) and a creep crack growth law of the form $da/dt = B.C^q$ where C represents a suitable crack tip parameter. Indeed its choice represents a particularly widely studied area; here attention is limited to two established procedures: C^* based analysis, which, coupled with reference stress concepts, forms the basis of the approach specified by the British R5 code (2,3), and C_t based analysis as developed in US by Saxena and co-workers (4,5) and implemented in the PCPIPE code (6). Both are methods distinguished by the fact that they offer well-defined procedures for evaluating the parameter for typical component crack geometries and loading regimes (in the case of R5 using handbook stress intensity solutions together with limit load analysis and for the C_t approach via analogy with the EPRI elasto-plastic fracture mechanics functions). The first step to preparing the way for their application remains the establishment of a material properties database, including stress-strain curves and creep behaviour laws in addition to the fracture mechanics parameters and laws mentioned above. The present paper reports firstly on testing carried out with this goal in mind (and as such forms part of a larger Italian project regarding the introduction of advanced high chromium steels for thermal power plant). The second section compares the predictions of the analysis methods with selected test data.

Tab. 1 Alloy composition (%wt.)

Element	P91	E911
C	0.09	0.105
Si	0.45	0.20
Mn	0.44	0.35
P	0.018	0.007
S	0.001	0.003
Cr	9.08	9.16
Mo	0.96	1.101
Ni	0.08	0.23
Nb	0.08	0.068
V	0.19	0.23
Al	0.005	0.007
W	-	1.00
N_2	0.041	0.072

Table 2 Tensile properties

		RT	300°C	600°C
P91	$\sigma_{y,0.2\%}$, MPa	505	432	278
	UTS, MPa	679	553	338
E911	$\sigma_{y,0.2\%}$, MPa	541	444	323
	UTS, MPa	722	582	404

2. EXPERIMENTAL

2.1 Materials and Welding

Two steels are considered: modified 9Cr-1Mo (P91) and E911. P91 is now being widely used for new steam plant and for retrofits; for this study specimens have been machined from two pipe sections (ϕ343x73 mm and ϕ545x38 mm respectively) such as would be used for headers and main steam lines. Both originate from a single Cogne cast and were forged by Dalmine. E911 is an advanced 9% chromium steel with 1% tungsten and is currently being characterised in the European COST-501 programme. The pipe reported here (ϕ545x38) was forged by Dalmine from a British Steel Engineering Steels ingot. The chemical composition and tensile characteristics of both materials are compared in Tabs. 1 and 2 respectively.

The P91 pipe was butt welded following a SAW automated procedure in the case of the ϕ343 mm pipe and a manual SMAW procedure for the ϕ545 mm pipe. In both bases matching Mod.9Cr1Mo fillers (Böhler) were used, together with a 760°C/4h PWHT.

2.2. Fracture Toughness Testing

Fracture toughness tests have been performed over the temperature range RT to 600°C according to ASTM E813-89 using C(T) specimens from the P91 and E911 base materials as well as the P91 welds. These "fusion line" or "HAZ" specimens were machined with the crack tip in the HAZ and with the crack plane orientated parallel to the fusion line (Fig. 1). This location was preferred over the weld metal itself because in high temperature service it is probable that damage will initiate in the HAZ. All the specimens were side grooved after pre-cracking, reducing the section area by 20%. Crack growth was monitored via a potential drop system. For the tests at temperatures of up to 280°C the load-line displacement was followed using a clip gauge inserted in the crack mouth, whereas for higher temperatures an indirect method based on the test system's compliance curve was applied.

All three materials show good toughness over the temperature range examined (Fig. 2). For P91 at RT, the average J_{Ic} value of 223 kJ/m^2 is almost twice that reported for mod.9Cr1Mo in the literature (7). E911 proved even tougher with a mean of 302 kJ/m^2; the J-R curves for the two materials are compared in Fig. 3. Also the values for the P91-HAZ specimens are in good agreement with the trend established for the base material i.e. the HAZ-weld microstructures appear to have no detrimental effect on the fracture toughness.

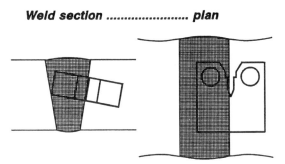

Fig 1 Orientation of the P91-HAZ (fusion line) specimens.

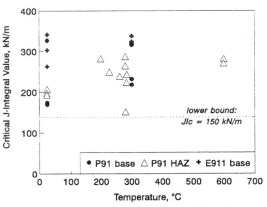

Fig 2 Temperature dependence of J_{Ic} values.

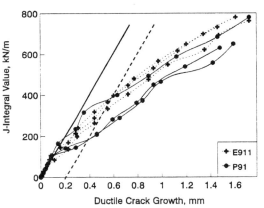

Fig 3 J-R curves for P91 and E911 at RT.

2.3 Fatigue Crack Growth (FCG)

FCG tests have been performed at 600°C and with R=0.1 on P91 parent, P91 fusion line and E911 parent specimens. Testing frequencies ranged from 20 Hz (standard fast sinusoidal cycling for which creep effects should be negligible) down to 0.01 Hz.

Under fast cycling conditions the crack growth rate - stress intensity data for all three specimen types proved very similar (Fig. 4) and in good agreement with literature results for P91 (7,8). Decreasing the cycling frequency produced a modest increase in crack growth rates for a given ΔK value for the P91 parent and fusion line specimens, an effect that appeared more pronounced at higher values of ΔK (Fig. 5). In contrast E911 proved insensitive to such effects for the frequencies/temperature values examined (20 → 0.1 Hz). It was noted that the P91 fusion line specimens produced results in good agreement with those of the parent material, despite the fact that the crack propagation path through the HAZ/weld microstructure varied from specimen to specimen (Fig. 6).

Post-test fractography showed no effect of frequency on the P91 fracture surfaces, which were entirely transgranular. On the other hand, microhardness measurements revealed the presence of significant softening ($\approx 12\%$) in a 0.2 mm band close to the fracture surface of the lowest frequency parent material specimen. In contrast that tested at 20 Hz showed no variation in hardness. This softening effect at low testing frequencies is attributed to the growth of an extended plastic/creep zone; in fact post-test hardness measurements on creep and LCF specimens typically indicate a 10-20% reduction in hardness due to a combination of strain-based and thermal softening.

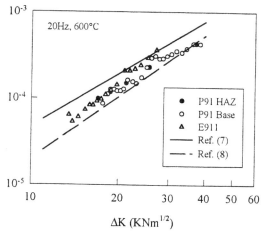

Fig 4 Fatigue crack growth data at 600°C, 20 Hz, R=0.1 for P91 and E911.

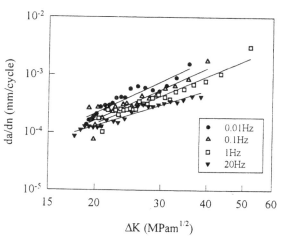

Fig 5 Frequency effect in FCG tests on P91 (600°C, R=0.1)

2.4 Creep Crack Growth

Two creep crack growth tests have been performed ay 600°C on C(T) fusion line specimens of thickness 12 mm. The load-line displacement was measured using a probe positioned in

(a) J_{Ic} specimen, 25°C, J_{Ic} = 208 kN/m: crack through the FG-HAZ

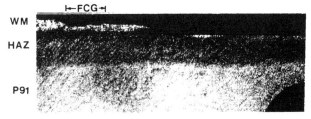

(b) FCG, 600°C, 20 Hz: pre-crack in CG-HAZ, crack then runs in WM

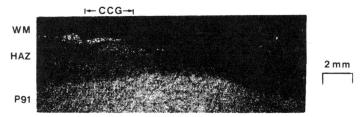

(c) CCG, 5 kN, 600°C: pre-crack in CG-HAZ, growth into the WM

Fig. 6 Fracture paths in P91-HAZ specimens (WM = weld metal, CG-HAZ = coarse-grained heat affected zone, FG-HAZ = fine-grained heat affected zone).

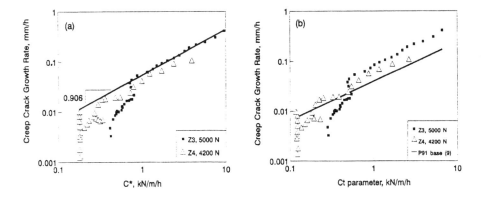

Fig. 7 Correlation of the P91-HAZ creep crack growth data (a) with C^* and (b) with C_t.

Table 3 Expressions used for calculating the crack tip parameters.

C^* using the R5 approach (valid for large scale secondary creep) $$C^* = \sigma_{ref} \cdot \dot{\epsilon}_{ref} \cdot R' \qquad R' = K^2/\sigma_{ref}^2 \qquad \sigma_{ref} = P \cdot \sigma_y / P_{Lc}$$
C^* using EPFM analogy (valid for large scale secondary creep) $$C^* = (W-a) \cdot A \cdot \sigma_{ref}^{n+1} \cdot h_1(a/w,n) \qquad \sigma_{ref} = \sigma_{net}/1.455\eta$$
$C^*(t)$ (valid for large scale primary creep - secondary creep (PC-SC)) $$C^*(t) = \frac{C_h^*(p,A_1,n_1,a/W,P)}{(1+p)\, t^{p/(1+p)}} + C_s^*$$
C_t (valid for small and large scale creep for secondary creep) $$C_t = 2(1-\nu^2)\frac{K^2}{E W}\frac{F'}{F} \beta \, \dot{r}_c + C^*$$ $$\dot{r}_c\big|_{SC} = 2\alpha/(n-1) \cdot K^2 \, t^{-(n-3)/(n-1)} \, (EA)^{2/(n-1)} \, \bar{r}_c(\theta)$$ C_t (valid for primary and secondary creep regimes) $$C_t = 2(1-\nu^2) \, K^2/(E W) \, F'/F \, \beta \, \dot{r}_c + C^*(t)$$ $$\dot{r}_c\big|_{PC-SC} = \frac{K^2}{2\pi}\left[\frac{I_{n1}E}{2\pi(1-\nu^2)}\right]^{\frac{2}{n_1-1}} [(1-n_1)(1+p)A_1]^{\frac{2}{(1+p)(n_1-1)}} \left[\frac{\bar{r}_c(\theta)}{1+p}\right]\left[\frac{2}{n_1-1}\right] t^{\frac{2}{(1+p)(n_1-1)}-1}$$

the crack mouth. Figs. 7a and b show the correlations obtained with the C^* and C_t parameters calculated on the basis of the experimental load-line and PD crack growth measurements. Metallographic examination of the mid-section of specimen tested at 5 kN ($K_o \approx 30$ MPa$\sqrt{m}$) indicated that although the pre-crack grew in the coarse-grained HAZ, during the test itself the crack deviated towards the fusion line and propagated through the weld metal.

3. ANALYSIS

While the above data can certainly be used for comparative purposes, their eventual application in life assessment studies is equally important. Given in particular the well-known sensitivity of creep crack growth analysis, verification of the envisaged procedures is essential, bearing in mind that if these are to be of value to the plant operator they need to be conservative, but not overly so. As a modest first step to assessing the accuracy of C^* and C_t approaches (Tab. 3), predictions can be compared with laboratory test results; here we consider the frequency effect noted in the FCG tests on P91 and the creep crack growth rupture times. The overall database established for P91 at 600°C is indicated in Tab. 4. Some

Table 4 Database for P91 crack growth analysis at 600°C

Cyclic σ-ϵ curve [mm/mm]:		
$\epsilon = \sigma/E + D\,(\sigma/\sigma_y)^m$	E	155000 MPa
	σ_y	218.5 MPa
	D	0.002
	m	13.48
Creep rate law [1/h]:	A_1	5.1×10^{-33} MPa^{-3n1}/h
	n_1	3.614
$\dot{\epsilon}_c = A_1 \sigma^{m(1+p)} \epsilon_c^p + A_2 \sigma^n$	p	2
	A	4.5×10^{-32} 1/h
	n	12.2
FCG: Paris Law [mm/N]		ΔK in MPa$\sqrt{m}$
	S	9.3×10^{-8}
$da/dN = F\,\Delta K^v$	v	2.412
CCG [mm/h]		C^* in kN/m/h
		q = 0.906; B = 0.0545
$\dfrac{da}{dt} = B\,(C_{param})^q$		C_t in kN/m/h
		q = 0.752; B = 0.042

of these relations have been calibrated within the present programme (the Ramberg-Osgood equation to describe the cyclic σ-ϵ curve, the Paris law for FCG, fracture toughness); in other cases the parameters have been obtained from the literature. In particular, the parameters in the primary creep constitutive equation have been derived from the expressions given by Jaske and Swindeman (9).

3.1 Creep-Fatigue Crack Growth Rates

The aim is to evaluate to what extent creep processes could be responsible for the increase in crack growth rate observed at low frequencies for the P91 base and fusion line specimens. Such an interaction is typically modelled using the relation:

$$\left.\frac{da}{dN}\right|_{tot} = \left.\frac{da}{dN}\right|_{fat} + \int_0^{t_{period}} \left.\frac{da}{dt}\right|_{creep} \quad (12)$$

The first term represents the pure fatigue component (from the relevant Paris law) while an appropriate creep crack growth law parameter is integrated over the loading cycle to estimate the creep contribution (the role of oxidation/environmental effects is not directly considered). Fig. 8a compares the experimental crack growth rate vs. frequency trend for P91 at an applied ΔK of 30 MPa$\sqrt{m}$ with curves predicted from eqn.(1) using the various C parameters reported in Tab. 3. For the maximum value of K during such a cycle the transition time to large scale creep is ≈ 260 secs if a primary creep model is applied and over 2½ hours for a pure secondary creep model. Therefore, given the relatively short duration (100 s.) of even the lowest frequency cycle, the C_t parameter would be expected to be the most appropriate since it was developed to account for small-scale creep effects. However the growth rates generated (using a secondary creep law) by such a model are found to be very high compared to the experimental values (introduction of primary creep coefficients effectively worsens the situation). Given the fact that the small-scale creep term of C_t depends inversely on a time parameter, the assumption that this is reset to zero at the start of each cycle appears to be overly conservative. On the other hand the C^* based prediction underestimates the frequency effect, as might be expected of a large scale creep model.

The same approach has been applied to the E911 steel on the assumption that the minimum creep rates are a factor of ≈ 5 lower than for P91 at a given stress and that the yield strength is 10% higher. As shown in Fig. 8b, the C^* prediction corresponds better with the experimental evidence, indicating no creep-fatigue interaction over the frequency range of the limited experimental data available.

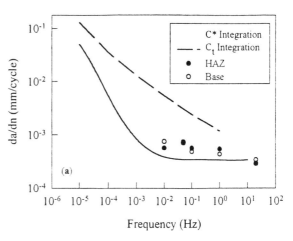

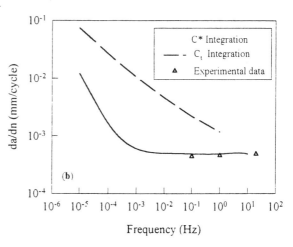

Fig. 8 Creep-fatigue crack growth interaction for (a) P91 and (b) E911, both at 600°C, $\Delta K = 30$ MPa$\sqrt{}$m.

3.2 Creep Crack Growth Life

CCG life predictions obtained by integration of the crack growth law have been compared with the failure times obtained in the two tests reported above to examine their sensitivity to crack tip parameter/material laws parameters. The variations considered are:

a) for steady state (Norton) creep: C^* as calculated by the R5 approach, C^* as calculated using the EPFM analogy and C_t, which includes consideration of small scale creep (the C^* expressions in contrast relate to large scale creep conditions);

b) for the primary-secondary creep law: C^* (EPFM analogy version) and C_t.

Table 5 Transition time values for the P91 CCG test conditions

Transition parameters		a/w=0.528, 5000 N	a/w=0.488, 4200 N
primary-secondary creep law	t_1, transition to large scale primary creep	0.13 hours	2.17 hours
	t_2, transition to large scale secondary creep	16.7 hours	21 312 hours
secondary creep law	t_T, transition to large scale secondary creep	5.1 hours	396 hours
Experimental rupture time		55 hours	226 hours

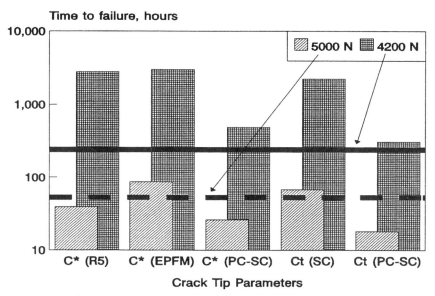

Fig. 9 Time to rupture predictions for the creep crack growth tests on P91 HAZ specimens at 600°C.

Fig. 9 compares the theoretical life values obtained (N.B. any incubation period was ignored; failure was defined as the reference stress exceeding the yield stress or J_{tot} exceeding J_{Ic}; finally, plain strain conditions were assumed). Although for the higher load test the secondary creep law based failure time values are in reasonable agreement with the experimental life, in the case of the 4.2 kN test these prove extremely non-conservative. Consideration of the primary creep behaviour (which acts to accelerate the initial crack growth rate and reduce the theoretical failure time) significantly improves the quality of the predictions. The effect of modelling the small scale creep regime also has a positive (albeit less significant) effect, as confirmed by the transition time values shown in Tab. 5.

4. CONCLUSIONS

The fracture toughness and high temperature fatigue crack growth rate values of P91-HAZ specimens are found to be comparable with those of the parent material; furthermore they appear relatively insensitive to the propagation path (i.e. through the HAZ, fusion line or weld metal). This may not be the case in serviced components in which the crack path would be expected to favour creep damaged regions, in particuar in the fine-grained HAZ.

The E911 (9Cr with %1 W) specimens produced comparable, if not better, toughness and FCG properties with respect to the modified 9Cr steel.

C^* and C_t-based analyses have been applied to predict frequency effects in FCG tests and failure times for creep crack growth specimens. For FCG conditions for P91, consideration of the small scale creep effects leads to unrealistically high crack growth rates for the test conditions simulated. The assumption that the time parameter in the C_t small-scale creep expression be set to zero at the start of each cycle appears overly pessimistic.

For creep crack growth it is found necessary to take account of the primary creep behaviour of P91 if realistic failure time predictions are to be obtained. However the sensitivity of the predictions to the analysis procedure and the input parameters underlines the need for further verification tests (in particular long term, low crack growth rate tests) if these methods are to be applied with confidence to components.

References

(1) Fontana, E., De Michelis, C. & Ghia, S., AE Trek: From First Hydraulic Test to Power Plant Component Monitoring, Progress in Acoustic Emission VII, Eds. Kishi et al, Japanese Society for NDI, 1994, pp.39-49.

(2) Webster, G. and Ainsworth, R.A., High Temperature Component Life Assessment, Chapman & Hall, 1994

(3) Ainsworth, R.A., Assessment of the high-temperature response of structures: developments in the R5 procedure, 6th Int. Conf. on Creep and Fatigue, IMechE, 1996

(4) Saxena, A., Mechanics and mechanisms of Creep Crack Growth, Fracture Mechanics: Microstructures and Mechanisms, Eds. Nair et al, ASM, 1987, pp.283-334

(5) Saxena, A., Fracture Mechanics Approaches for Characterising Creep-Fatigue Crack Growth, JSME Int. Journal, A, Vol. 36, No. 1, 1993

(6) Liaw, P.K., Saxena, A. and Schaefer, J., Predicting the Life of High-Temperature Structural Components in Power Plants, JOM, February 1992, pp. 43-48.

(7) James, L.A. and Carlson, K.W., The Fatigue Crack Growth and Ductile Fracture Toughness Behaviour of ASTM A387 Grade 91 Steel, ASME J. Pressure Vessel Technology, 1985, Vol. 107, pp. 271-278

(8) Vanderschaege, A., Gabrel-Cousaert, C., Lecoq, J., Modified 9% Cr Steel (Grade P9.1): Low Cycle Fatigue and Crack Propagation Rate of Base Material and Welds at Room and High Temperature, Int. Conf. Pressure Vessel Technology - 7, 1993.

(9) Jaske. C.E. and Swindeman, R.W., Long-Term-Creep and Creep-Crack-Growth Behaviour of 9Cr-1Mo-V-Nb Steel, Advances in Materials Technology for Fossil Power Plants, ASM, 1987, pp. 251-258.

Emerging Technologies

C522/031/97

Emerging coal-fired power generation technologies

D H SCOTT BSc, FCOMA, MIMgt
IEA Coal Research, London, UK

ABSTRACT

Developments in coal-fired power generation technology are being driven by the need to minimise the costs of electricity production while complying with increasingly strict emissions regulations. Several new systems are available but power project developers are understandably reluctant to commit major investment to unproven technology. Large scale projects now underway may help to establish the required confidence.

1. INTRODUCTION

The processes of industrialisation and urbanisation are inevitably accompanied by a large increase in the demand for electric power. The benefits of a ready supply of economically priced electricity are widely appreciated but the problems arising from the combustion of fossil fuels have also become apparent. The effects of 'acid rain' and photochemical smog have led to increasing emphasis on pollution control in the developed countries and the need for emissions control in developing countries becomes more apparent as energy consumption increases. There is also concern that burning fossil fuels is increasing the concentration of CO_2 in the atmosphere. Paradoxically, the consequence of adding pollution controls to power stations is to reduce their overall efficiency and hence to increase CO_2 emissions. The only practical strategy for reducing specific CO_2 production (grams of CO_2/kWh) is by increasing efficiency. Hence, there is a need to develop cleaner, more efficient means for satisfying the increasing demand for electricity. Where economically priced gas is available, low investment costs, high efficiency and the relative simplicity of emissions control make combined cycle gas turbine the system of choice. However, coal fired generating capacity will continue to increase because it is the only near term option for achieving the enormous increase in power required in India and Asia.

A range of new coal fired technologies is available to utilities but, in this context, the term 'available' has to be qualified by commercial considerations such as the cost, complexity and reliability of the technology. The cost of a new 500 MWe power station is around £250 million. For a project of this scale the choice of an inappropriate technology could be a financial disaster. It is not surprising that recent surveys have shown that the developers of power supply projects, and their funding agencies, are conservative in their choice of power generating technologies (1).

2. THE EVOLUTION OF CONVENTIONAL PULVERISED COAL FIRED POWER STATIONS.

During the first 60 years of the 20th century the technology for building and operating fossil fuel fired power stations developed rapidly. Thermal efficiency, on a higher heating value basis, rose from less than 5% to around 35%. By the mid 1960s efficiency had levelled off at around 40% for the best plants and 33% for average plants (2). Much of the increase in efficiency was achieved by increasing the main steam temperatures and pressures. The most efficient power stations use supercritical conditions where the boiler water is heated at a pressure in excess of its critical pressure (22.1 MPa). The first supercritical power stations were constructed at the end of the 1950s. The US power plant Eddystone 1 was commissioned in 1958 and had design conditions of: 34.4 MPa main steam pressure, 649^0C main steam temperature and two stages of reheat each to 566^0C (34.4 MPa/649^0C/566^0C/566^0C). The need for high creep resistance under these conditions led to the use of thick section austenitic stainless steels for items such as the boiler tubes, main steam pipelines and valves. The design efficiency was 43% but, because of boiler tube failures, the station had to be derated giving an efficiency of 41% (3). Problems with the first generation of supercritical boilers led to the conclusion that pulverised coal fired electricity generation was a mature technology with an efficiency limited by practical and economic considerations to around 40%. Commercially the subcritical cycle with one stage of reheating (16.6 MPa/538^0C/538^0C) is still the dominant design.

3. ENVIRONMENTAL CONSTRAINTS ON POWER GENERATION

Concern about the effects of air pollution has led to the establishment of air quality standards and to comprehensive surveys identifying the sources of air pollutants. Corinair, a part of the 1985-1990 EC 'CORINE' research programme furnished the results for aggregate emission levels of various pollutants in 29 European countries (the 15 countries of the EU plus Norway, Switzerland, Bulgaria, Croatia, Czech Republic, Estonia, Hungary, Latvia, Lithuania, Malta, Poland, Romania, Slovakia, and Slovenia). The data showed that from total annual emissions of 28 million tonnes of SO_2, 18 million tonnes of NO_x and 4,765 million tonnes of CO_2, public power, cogeneration and district heating contributed 15 million, 3.8 million and 1,332 million tonnes, respectively.

At the end of 1994 the European Commission (EC) asked the European Environmental Agency (EEA) to prepare a state of the environment report for the European Union. The main conclusions of the EEA report are that the European Union is making progress towards reducing certain pressures on the environment but this is not enough to improve the general quality of the environment and represents even less progress towards the declared aim of sustainability (4). In response the EC is reported to have drafted an action plan promising that a broader mix of instruments will be used to attain environmental targets. These include more effective enforcement, the application of environmental liability, environmental charges, fiscal reform and a framework for voluntary agreements (5). Although a European example has been quoted here a similar process of evaluation and regulation is taking place at many other locations around the world.

An 'uncontrolled' PC fired power station burning 2.5% sulphur coal would release flue gas containing around 4700 mg/m^3 of SO_2, 800 - 2000 mg/m^3 of NO_x and around 8g/m^3 of dust. Generally, new power stations are obliged to observe emission limits of 400 mg/m^3 of SO_2, 650 mg/m^3 of NO_x and 50 mg/m^3 of dust. Where air quality is a sensitive issue more rigorous local requirements may apply. For example, some countries specify a maximum coal sulphur content of 1% and a minimum sulphur removal from the flue gas of 85%. Local NO_x emission limits of 200 mg/m^3 are widely applied and dust emission limits may be as low as 10 mg/m^3. It appears that the PC fired power station of the future may need flue gas desulphurisation of 95% efficiency or better, selective catalytic reduction or equivalent technology to control NO_x emissions, and possibly a final wet electrostatic precipitator or baghouse to control particulate emissions. This will add to the costs of power generation and, unless further increases in the basic efficiency of the boiler/turbogenerator set can be achieved, the parasitic power consumption of the additional equipment will increase specific CO_2 emissions.

4. NEW POWER GENERATING TECHNOLOGIES

Currently the improved technologies showing the most early promise for commercial development appear to be: supercritical and ultrasupercritical PC, atmospheric pressure circulating fluidised bed combustion (CFBC), pressurised fluidised bed combustion (PFBC) and integrated gasification combined cycle (IGCC).

4.1. Supercritical and ultrasupercritical PC

In spite of the disappointing results from the first generation of large supercritical boilers, research and development of the technology has continued. The first generation used high alloy austenitic stainless steel for the boiler tubes and for headers and piping outside the boiler. The furnace and convective sections of modern boilers are contained by continuous membrane walls that are fabricated from low alloy steel. Careful design and control ensures that the metal is not overheated during normal operation. High pressure piping and headers outside the boiler are made of ferritic stainless steel. The use of these materials, which are less susceptible to stresses induced by temperature changes, has facilitated the design of a new generation of supercritical boilers suitable for load following and/or daily stop-start operation. The use of supercritical boilers has extended beyond the USA, Europe and Japan and it is now possible to describe supercritical boilers having steam conditions of 25 MPa/540^{0}C/560^{0}C as conventional. The first supercritical boilers in China, two 600 MWe units at Shidongkou near Shanghai, were commissioned in 1992. The electrical efficiency of the Chinese units is quoted as 39.7% (6). Korea Electric Power Corporation is in the process of installing twenty 500 MWe supercritical boilers. Six were in operation by the end of 1996 and the remaining 14 are planned to be working by the year 2000 (7).

Further advances to 'ultrasupercritical' steam conditions have been made possible by the availability of a high strength ferritic steel (ASTM T91/P91) that was specially developed as part of the US breeder reactor programme. The use of this material for fossil fuel fired boilers was first demonstrated in 1989 at Kawagoe, a LNG fired power station with a main steam pressure of 32.5 MPa and a main steam temperature of 571^{0}C. Commissioning of the Nordjyllands power station in

Denmark is scheduled for 1998. This coal fired ultrasupercritical power station with highly optimised steam conditions and access to exceptionally cold cooling water has a designed efficiency of 45%. Further development of ultrasupercritical technology will depend on the availability of superior steels but it is currently envisaged that a coal-fired power station with an efficiency of 48% might be in operation by 2005 with over 50% possible by 2015 (8).

It is generally considered that there is a capital cost premium associated with supercritical boilers that limits their commercial application to areas where high fuel cost provides the necessary incentive for increased thermal efficiency. However, developments in construction techniques and designing for optimum rather than maximum efficiency may eliminate or reverse the differential (9, 10, 11).

4.2. Atmospheric fluidised bed boilers

Circulating fluidised bed combustion (CFBC) boilers were developed in the mid 1970s (*see* Figure 1).

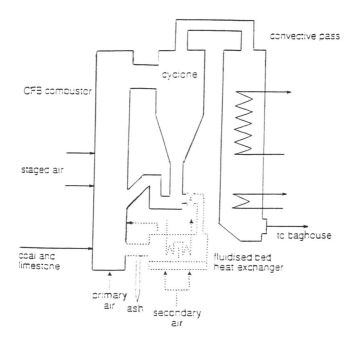

Figure 1 Circulating fluidised bed boiler (12)

A mixture of crushed coal and limestone is injected into a bed of ash which is fluidised by the primary air. The fine pulverisation required for PC firing is unnecessary, indeed undesirable, for CFBC. A high fluidising velocity (5-10 m/s) is used to ensure that a substantial proportion of the bed material is carried over with the combustion gases. This material is collected in the cyclone and recycled to the fluidised bed. Recirculation of the coal and limestone particles ensures good gas/solids contact. CFBC boilers are capable of achieving relatively low levels of the primary pollutants SO_2 and NO_x without the need for expensive pollution control equipment. SO_2 emissions are controlled in-situ by the added limestone. The low combustion temperature (800-900^{0}C) limits the formation of NO_x. Despite these low temperatures, CO and unburnt hydrocarbon emissions are also low because of good solids and gas mixing and long residence times in the bed. CFBC boilers are ideally suited to high ash, low sulphur coal. They can accommodate high sulphur coals but achieving low sulphur emissions requires increased limestone addition with consequent increases in procurement and disposal costs. The requirement for flue gas desulphurisation imposes a considerable capital and operating cost penalty on a small grate fired or PC fired boiler. In consequence, CFBC boilers have achieved considerable success as relatively small units (~80 MWt) exploiting low value or waste fuels such as washery discards. For bituminous coals of international trade quality, a CFBC boiler generally needs more heat exchange surface than the equivalent PC boiler and has a higher 'in house' energy demand. The large cyclones also tend to increase heat losses. In consequence, its efficiency tends to be marginally less than a PC boiler with similar steam conditions and, with increasing unit size, economies of scale tend to cancel CFBC's initial cost advantage. However, if utility scale CFBC boilers were available they might be commercially competitive at locations where there is an assured supply of low cost, low rank or low grade coal. In India for example, the coal used in power stations is typically low sulphur (0.2-0.3%) but contains around 40% ash and 15% water. The ash has a high quartz content which, with the moisture content, contributes to the rapid wear of coal pulverisers (12). The ash is also reported to have a calcium content of around 1-2% CaO (13). If the calcium is in a suitable form it might provide a 'free' sulphur sorbent.

Although Indian coals appear ideally suited to CFBC, Western project developers still specify multiple 200-600 MWe PC boilers for Indian power projects. The lack of experience in the +200 MWe range is undoubtedly militating against CFBC deployment (14). This problem may be mitigated in the next few years as the large CFBC boilers currently coming into operation increase confidence in the technology. The largest CFBC boiler in operation is at Gardanne in France. Commissioning was completed in October 1995. This 250 MWe single reheat subcritical boiler (16.3 MPa/565^{0}C/565^{0}C) burns a local, high sulphur (3.7%) subbituminous coal containing 28-32% ash. Although the sulphur content is high, 57% of the ash is CaO and, with the addition of some limestone from mine waste, 97% SO_2 removal is achieved (15). A 200 MWe CFBC boiler is currently being built in Korea. The unit is designed to use the local low grade anthracite (ash 40-45% sulphur 0.6-1.5%) and hence to reduce Korea's dependence on imported coal. Initial firing of the boiler is scheduled for Autumn 1997 with commercial operation in Summer 1998 (16).

Currently all the CFBC boilers in operation feature subcritical steam conditions. A recent study indicates that there are features of CFBC boiler design that make it particularly suitable for supercritical operation (17):
* Low heat flux; the peak heat input per unit area of heat exchange surface of a CFBC boiler is much less than that of a PC fired boiler. The heat flux in the middle region of a PC boiler, immediately above the burners, is of the order of 270 kW/m^2 (18). Skowyra and others (17) give a value of 90 kW/m^2 for a CFBC boiler.
* Cleaner combustor wall; compared with a PC fired boiler there is very little deposition on the walls.
* Heat flux profile; the peak heat flux in a CFBC combustor is near the bottom and gradually decreases as the height increases. Water enters at the bottom of the boiler and hence is coolest where the heat flux is greatest. This arrangement helps to minimise wall tube metal temperatures.

4.3. Pressurised fluidised bed combustion

Currently the most efficient steam cycles use main steam temperatures approaching 600^0C. The bed temperature of a fluidised bed boiler is around 850^0C. Potentially, a cycle with this upper temperature could be more efficient than available steam cycles. These considerations have led to the design of pressurised fluidised bed combustion (PFBC) systems in which some of the exergy in the hot gas leaving the combustor is exploited directly using an expansion turbine. The only commercially available PFBC systems are based on ABB-Carbon's pressurised bubbling fluidised bed technology. The essential features of the process are shown in Figure 2.

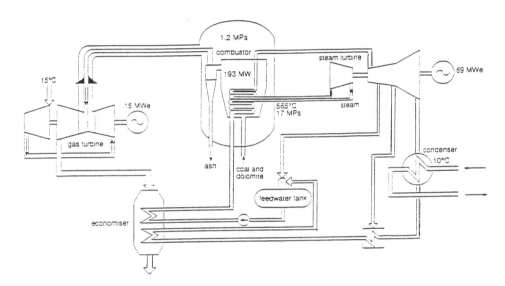

Figure 2 ABB PFBC P200 unit (19)

The PFBC process has some inherent advantages over both PC and atmospheric FBC. Among the advantages are:
* a PFBC unit is more compact than an atmospheric pressure unit or a PC boiler of comparable output. Thus PFBC could be particularly suitable for repowering operations where an existing boiler in a constricted site needs to be replaced by a cleaner, more efficient system;
* the PFBC boiler is essentially part of a combined cycle system. The efficiency is strongly affected by the efficiency of the steam cycle but the expander provides a useful bonus. Generally, the PFBC system with a subcritical steam cycle gives an efficiency similar to that from basic supercritical PC (~40%) while a PFBC system with a supercritical steam cycle is expected to give an efficiency around 42%.
* as with atmospheric pressure CFBC, in bed sulphur absorption is achieved by injecting limestone or dolomite. Because of the effects of pressure PFBC gives greater sulphur capture efficiency at lower Ca/S ratios.

The first commercial PFBC boilers were installed at Stockholm Energi's Värtan site in Sweden. The site supplies the largest of the four district heating systems which serve Stockholm. Two ABB P200 modules serve a single steam turbine to provide 135 MWe and 225 MWt for district heating. The maximum heat demand is just over 1000 MW. Other systems on the site contribute to a total thermal capacity in excess of 1800 MW. The site dates from 1903 when it was outside Stockholm but the city has now engulfed it and the nearest dwellings are about 50 metres outside the perimeter fence. Because of its location the plant is subjected to unusually stringent environmental regulations. Initially it was required to meet a SO_2 emission limit of 160 mg/m^3 and a NO_x limit of 130 mg/m^3. These limits were subsequently tightened and the plant, which uses 1% sulphur coal, is now required to comply with a SO_2 limit of 50 mg/m^3 (desulphurisation efficiency >97%).

Following the commissioning of the Värtan plant PFBC plants based on the P200 module have been built in Spain (Escatron), the USA (Tidd), and Japan (Wakamatsu). A 360 MWe unit based on ABB's larger P800 system and using supercritical steam conditions is now nearing completion at Karita in Japan. The unit will use the ABB manufactured hot gas expander but most of the plant will be manufactured in Japan under ABB license. A combined heat and power plant based on the P200 is planned for Cottbus in Germany and further plants are planned for Japan. It is expected that PFBC will be considered fully commercial in Japan by the year 2005. The likely application will be for the repowering of existing 300-400 MWe PC boilers (20).

4.4. Integrated gasification combined cycle (IGCC)

Coal gasifiers are used in many countries for the commercial production of gas and chemicals. The high efficiency and clean operation of natural gas fired combined cycle plants has led to their use by an increasing number of utilities and the conversion of coal into a clean fuel gas has been proposed as the route to clean and efficient coal based electricity generation. Because potential pollutants, sulphur, nitrogen and ash, are removed in concentrated form before combustion the technology is considered to be inherently more suited to clean operation. IGCC has not yet been proved for commercial use but currently there are five major IGCC demonstration projects, three in the USA and two in Europe:
* The 262 MWe, Wabash River, coal gasification, repowering project features an oxygen blown, entrained flow, slagging gasifier based on Dow gasification technology. The designed efficiency is 38% using locally mined high sulphur coal;
* The 260 MWe Tampa Electric project will feature Texaco's oxygen blown entrained flow slagging gasifier. The design efficiency of this unit is 39%;
* The Piñon Pine gasifier project features a Kellog Rust Westinghouse pressurised fluidised bed gasifier. Rated output is 99 MWe;
* Demkolec is operating a 250 MWe plant at Buggenum in the Netherlands based on the Shell, entrained flow, oxygen blown, slagging gasifier;
* Elcogas in Spain are preparing to demonstrate their Prenflow oxygen blown slagging gasifier in conjunction with an advanced gas turbine.

The Spanish plant with a capacity of 335 MWe will be the largest IGCC plant and is expected to have an efficiency of 43%. Anticipated atmospheric emissions are: SO_2 <25 mg/m^3, NO_x <150 mg/m^3 and particulates < 7.5 mg/m^3. Commissioning is scheduled for 1997 and there will be a demonstration period of three years for testing various fuels and technology improvements.

5. CONCLUSIONS

Power project developers are understandably conservative when committing the considerable investment needed for a major power project. Subcritical PC is regarded as a fully proven technology. Supercritical PC, with a considerable number of fully commercial plants widely distributed around the world, appears to be approaching similar status. Ultrasupercritical PC, and utility scale CFBC and PFBC are at an earlier stage of development and the commercial deployment of IGCC may be still further in the future. CFBC appears to be particularly promising for 'difficult' coals and may make an important contribution to the environment of developing countries. PFBC has the advantages of higher efficiency than CFBC and a more compact boiler. IGCC may offer the ultimate in environmental friendliness but PC, CFBC, PFBC and IGCC are all able to comply with existing environmental requirements and have scope for further improvement.

6 REFERENCES

(1) **OECD/IEA (1996)** *Factors affecting the take-up of clean coal technologies - overview report.* Paris, France, OECD/IEA, 73 pp (1996)

(2) **Hirsch R F (1990)** New technologies for the utility industry: an historical and strategic perspective. In: *Power-Gen 89,* New Orleans, LA, USA, 5-7 Dec 1989. Houston, TX, USA, Power-Gen, pp 133-148 (1990)

(3) **Pace S, Sopocy D, Stenzel W (1994)** Development of an economic advanced pulverized coal utility plant commercial design. In: *Proceedings of the 19th international technical conference on coal utilization and fuel systems,* Clearwater, FL, USA, 21-24 Mar 1994. Washington, DC, USA, Coal and Slurry Technology Association, pp 253-264 (1994)

(4) **EEA (1996)** *EEA annual report 1995.* Luxembourg, Office for Official Publications of the European Communities, 55pp (1996)

(5) **Grant I (ed) (1996)** Action plan to strengthen enviro-policy. *Environment Business,* 3 (Jan 1996)

(6) **Doran W G, Koutsoukos A, Vacek M G (1993)** Huaneng International Power Development Corporation Shidonkou second power plant. In: *PowerGen Asia '93,* Singapore, 13-15 Sep 1993. Singapore, Times Conferences and Exhibitions PTE Ltd., vol 2, pp 389-399 (1993)

(7) **Kjaer S, Boisen A D (1996)** The advanced pulverised coal-fired power plants - status and future. In: *PowerGen Europe '96,* Budapest, Hungary, 26-28 Jun 1996. Utrecht, the Netherlands, PennWell Conferences and Exhibitions, vol 2, pp 535-547 (1996)

(8) **Sormani F, Monti M A, Maffei E (1996)** Comparison between cost of electricity generated from coal by ultrasupercritical steam plants and IGCC. In: *PowerGen '96 International,* Orlando, FL, USA, 4-6 Dec 1996. Houston, TX, USA, PennWell Conferences and Exhibitions, vol 3, pp 223-234 (1996)

(9) **Kaneko S, Hashimoto T, Maruta T (1996)** 1,000 MW coal-fired supercritical sliding pressure operation boiler with vertical furnace waterwall. In: *PowerGen Europe '96,* Budapest, Hungary, 26-28 Jun 1996. Utrecht, the Netherlands, PennWell Conferences and Exhibitions, vol 2, pp 550-562 (1996)

(10) **Franke J, Cossmann R, Huschauer H (1995)** Benson® steam generator with vertically-tubed furnace. *VGB Kraftswerkstechnik;* **75** (4); 321-327 (1995)

(11) **Scott D H (1995)** *Coal pulverisers - performance and safety* IEACR/79, London, UK, IEA Coal Research, 83 pp (1995)

(12) **Boyd T J, Cichanowicz J E, Tavoulareas S (1989)** Integrated features of AFBC technology. In: *Fourth symposium on integrated environmental control*, Washington, DC, USA, 2-4 Mar 1989. EPRI-GC-6519, Palo Alto, CA, USA, Electric Power Research Institute, pp 19-22 (Sep 1989)

(13) **Sanyal A, Roy C (1996)** Impact of coal quality on design, operation and emission characteristics of Indian utility boilers. In: *PowerGen '96 International*, Orlando, FL, USA, 4-6 Dec 1996. Houston, TX, USA, PennWell Conferences and Exhibitions, vol 1, pp 298-312 (1996)

(14) **Simbeck D R, Johnson H E, Wilhelm D J (1994)** The fluid bed market: status trends and outlook. In: *10th Annual fluidized-bed conference*. Jacksonville, FL, USA, 14-15 Nov 1994. Burke, VA, USA Council of Industrial Boiler Owners, pp 35-47 (1994)

(15) **Delot P, Roulet V (1996)** Provence 250 MWe unit: the largest CFB boiler in operation. In: *PowerGen Europe '96*, Budapest, Hungary, 26-28 Jun 1996. Utrecht, the Netherlands, PennWell Conferences and Exhibitions, vol 2, pp 473-484 (1996)

(16) **Maitland J E, Schaker Y (1996)** Design of the 200 MWe (net) Tonghae Thermal Power Plant circulating fluidized bed steam generator. In: *PowerGen '96 International*, Orlando, FL, USA, 4-6 Dec 1996. Houston, TX, USA, PennWell Conferences and Exhibitions, vol 3, pp 83-92 (1996)

(17) **Skowyra R S, Czarnecki T S, Sun C Y, Palkes M (1995)** Design of a supercritical sliding pressure circulating fluidized bed boiler with vertical waterwalls. In: *13th international conference on fluidised bed combustion*, Orlando, FL, USA, 7-10 May 1995. New York, NY, USA, The American Society of Mechanical Engineers, vol 1, pp 17-25 (1995)

(18) **Stultz C S, Kitto J B (eds) (1992)** Fuel ash effects on boiler design and operation. In: *Steam: its generation and use*. Baberton, OH, USA, Babcock and Wilcox Co., pp 20/1-20/27 (1992)

(19) **Pillai K K, Wickström B, Tjellander G (1989)** The influence of gas turbine technology on the PFBC 200 power plant. In: *Second biennial PFBC power plants utility conference*, Milwaukee, WI, USA, 19-20 Jun 1986. EPRI-GS-6478, Palo Alto, CA, USA, EPRI Research Reports Center, pp 3/1/1-3/1/14 (Aug 1989)

(20) **Takahashi M, Nakabayashi Y, Kimura N (1995)** EPDC's update on Takehara 350 MW AFBC and Wakamatsu 71 MW PFBC demonstration tests and advanced power generation. *VGB Kraftswerkstechnik*; **75** (5); 390-394 (1995)

Plant Operation

C522/024/97

Contributions of by-pass systems to the flexibility of advanced steam plants

R R ROHNER Dipl Ing, ETH, U R BLUMER, and J AREGGER Dipl Ing, HTL
Sulzer Thermtec Limited, Winterthur, Switzerland

Bypass systems contribute to the operational flexibility of advanced steam plants. They allows short and reliably repeatable start-up times after planned and unplanned outages by minimising temperature gradients and mismatches on boiler and turbine. This becomes of the more importance the higher the operating pressure and temperature are.
The main criteria's for sizing a bypass system are discussed.
Flexibility is not only an operational aspect but also on aspect of plant availability and freedom of decision when components are damaged or when operating problems on subsystems arise. On two examples it is shown that the bypass manufacturer needs high technical competence and system knowledge to assist the plant operator effectively and quickly on solving such problems.

1. INTRODUCTION

From modern power stations, not only high efficiencies are requested, but also high life expectancies of the components and a high operational reliability and flexibility. The turbine bypass can contribute in fulfilling these requirements:

- It helps to minimise temperature steps and gradients on start-up and load rejection, i.e. to minimise stress and strain and therefore to prolong the life of the components.
- It allows short and reliably repeatable start-up times from any boiler and turbine condition, i.e. higher operational flexibility and lower energy losses on warm starts.
- It immediately takes over the boiler load on load rejection of the turbine/generator. In case of a grid failure the bypass-system allows the run-back to house load. In case of a turbine trip it allows to keep the boiler running. In both cases it allows to restore energy supply to the grid within shortest possible time i.e. the bypass helps to increase the time-availability of the plant.

These factors and subsequent correctly sized turbine bypass systems become of the more importance the higher the operating pressure and temperature are. Latest bypass specifications prove that 305 bar and 650 °C are state of the art for once through boilers.

This importance is even greater if cycling operation is planned. On the other hand it must not be forgotten that also on base load units the number of hot/warm starts is much higher than the number of cold starts i.e. a bypass system is of considerable help.

Practice has shown that often base load plants are switched to load cycling after some 10 or 15 years, because there are new generating systems with higher efficiencies.

By turbine bypass system we understand the arrangement shown in Fig. 1:

The steam is led from superheater outlet via the HP-Bypass into the reheater, from there via the LP-Bypass into the condenser, i.e. the HP-section as well as the IP/LP section of the turbine are bypassed.

2. ADVANTAGES AND SIZING OF THE BYPASS SYSTEM

2.1. The Bypass System during Start Up

The speed with which a unit can be started-up is limited by the given permissible life consumption, in other words by the allowable thermal stresses on boiler and turbine.

Fig. 2 is an example that shows the relation between steam-to-metal temperature mismatch, start-up time and life consumption.

For the plant flexibility the temperature match on warm and hot starts stands in the foreground. And this as mentioned in the introduction not only for cycling units. A correctly sized bypass-system allows the enthalpy required to be attained for the temperature match from every starting condition of boiler and turbine.

The following facts make reliably repeatable procedures possible:
- Firing rate and steam flow can be varied to control pressure and temperature. Without a bypass there would be a strong interdependence of these parameters.
- Relatively high flows through superheater (SH) and reheater (RH) allow high firing rates without endangering them by excessive metal temperatures.That means the boiler can be operated in a load range where the steam temperature follows the expected temperature vs. load curve.On very low steam flows the SH metal temperature is dictated by the exhaust gas temperature.
- The cold RH inlet temperature is controlled by water injection. This helps also the RH outlet temperature to be kept under control.
- With an adequately selected bypass system there are no restrictions concerning the attainable SH outlet temperature.

In addition to the good and easily attainable temperature match the bypass system offers some more advantages:
- The boiler can be operated on steady state conditions. This gives the possibility to schedule hold points, e.g. to bring in the coal mills, to accommodate auxiliaries more easily, to have time to clear minor faults and disturbances without interrupting the boiler operation.
- The high firing rate which is possible allows to start coal firing earlier, i.e. to save expensive oil.

- Once the conditions for turbine match are achieved the boiler can be operated at stable conditions with sufficient steam flow for a quick and reliable run-up and synchronisation of the turbine/generator.
- The automatic bypass control system allows the operator to pay his full attention to the turbine during this critical phase, without the necessity of corrective actions on the boiler side.
- The automatic bypass system is always ready to take over any unexpected disturbances during start-up.
- Practice has also shown that the start-ups of coal fired boilers can be made with firing rates where the heat absorbtion in the boiler is less sensitive to slagging and fouling. Poor heat absorbtion of the evaporater due to slagging and fouling would cause too high exhaust gas temperature at low firing rates and increase the risk of excessive superheater metal temperatures.
- Mainly from American literature it is known that start-up with a bypass system can help to minimise turbine blade erosion. Due to exfoliation of iron-oxide in the boiler during the start-up phase the steam contains highly abrasive impurifications. If during this period this steam is led through the bypass instead through the turbine then only a bypass valve stem would be eroded instead of far more expensive turbine blading.

Many of these advantages concerning flexibility and preservation of material are not only valid for hot and warm starts but also for cold starts.

Particularly on coal fired units where widely varying fuel qualities could cause heavy disturbances of the firing rate, with the subsequent effects to pressure and temperature, the bypass can help to keep boiler pressure in a narrow band.

The advantages that a turbine bypass system offers during the commissioning phase of a power station are obvious. Boiler trials that are usual when commissioning the firing system, can be carried out without stressing the turbine unnecessarily.

2.2 Sizing the Bypass System for Start-up

The bypass system as a help for start-up offers most advantages if it is designed for warm/hot starts after 6-10 hours standstill. If correctly sized for these conditions the bypass allows
- quick loading of the boiler independent of the turbine until the desired turbine match conditions are achieved,
- run-up and loading of the turbine up to bypass-flow within minutes.

That means achievement of shortest start-up time with optimum match for the hot turbine, i.e. with lowest stresses and lowest consumption of cycling life.

To select the correct HP-bypass size, the characteristics of turbine and boiler have to be considered in the following way:

The turbine metal temperature after the assumed time of standstill has to be estimated. As an example it is assumed that after a 6 hours shutdown this temperature would be 500°C. From the enthalpy/entropy diagram can be found that the necessary enthalpy to match this temperature after throttling in the turbine inlet valve is approx. 3490 kj/kg. (Fig. 3).

The diagram also shows that the SH outlet steam temperature is given by the boiler pressure selected for the start-up. The first two columns of Table 1 show the values for the assumed example. The SH temperature/flow characteristics now indicates the minimum SH-flow (or boiler load) that has to be set to achieve the requested SH outlet temperature. (Fig. 4).

Column 3 in Table 1 shows the resulting values for the selected typical example. In Fig. 5 these values are plotted in the pressure vs. flow diagram for the HP-bypass with the max. capacity as variable parameter. The nominal boiler outlet pressure is assumed to be 250 bar.

It can easily be seen that for a restart with 80 bar SH outlet pressure (Point 3) a bypass capacity of min. 80% MCR is required to match turbine temperature.

This example shows a simplified procedure. In practice it must be taken into account that the SH flow/temperature characteristic depends also on SH pressure.

There are also other considerations that might overrule this procedure:
- Boilers operating in sliding pressure mode need 100% HP bypass capacity to operate the boiler without the turbine on the sliding pressure line.
- For fast and reliable start-up of the turbine it is of advantage to have the pulverisers already in operation. In this case the minimum load on coal firing might become the sizing criteria.

For sizing the LP bypass valves two parameters have to be considered:
- The required LP bypass flow during start-up is given by the sum of the HP bypass flow and the flow of spraywater required for achieving the RH inlet temperature. As a result the LP bypass flow is approx. 15-18% higher than the HP bypass flow. Possibly a certain amount of auxiliary steam taken from the cold RH side can be deducted. (For the further discussion 100% LP-Bypass capacity is defined a 100% Boiler-MCR+HP-bypass spray flow at nominal RH-pressure.)
- The second parameter for determining the valve size is the RH pressure. When assuming that the IP/LP-intercept valves are closed then the RH pressure must be selected so that overheating of HP-turbine blading by rotation loss is avoided. In our experience the turbine supplier prescribes that in this phase the RH pressure must not exceed 25-30% of full load RH pressure.

In Fig. 6 the 4 assumed start-up flows for the HP-bypass example were transferred to the pressure vs. LP bypass flow chart (including HP spraywater).

It can be seen that for usual warmstart conditions a LP-bypass size of 80-100% MCR (incl. HP-spray) is required. Otherwise HP flow would have to be reduced for hot/warm-start with all the consequences such as low firing rates, mismatch of turbine inlet conditions as discussed before. Smaller capacities might even make a cold start more difficult.

2.3 The Bypass System when Normal Plant Operation is disturbed

Flexibility means also that after disturbances the supply of energy to the grid is restored as quickly as possible.

With disturbances on the boiler side, such as failure of feed pumps, fans or parts of the firing system, the steam flow produced by the boiler is insufficient and a bypass system cannot help in such cases.

With disturbances on the turbine side, two important cases have to be considered:
- Load rejection due to disturbance on the turbine/generator
- Load rejection due to loss of the grid.

In both cases there is immediately a high overproduction of steam. For units without bypass-system this causes - as a rule - a Main Fuel Trip (MFT). This is because the safety valves are blowing to the atmosphere, i.e. the RH has no flow and therefore excessive metal temperature would result immediately.

On coal fired units the consequence of a MFT is always a delayed restart of the unit because of the long procedures for the restart of the pulverisers.

Restart needs not only an inert atmosphere in many pulveriser types, but often it means also clearing the pulverisers from remaining fuel and this in many cases has to be done manually.

With a correctly sized turbine bypass system the heavy disturbances on pressure and temperature and long restarting times can be avoided.

The bypass system immediately takes over the excessive steam production: In case of loss of the grid the unit can run back to house load, in case of turbine/generator troubles the boiler can be kept running and brought back to an optimum waiting load.

When the disturbance is cleared the unit can be reloaded quickly or a safe decision to shut down the unit is possible.

2.4 Sizing the Bypass System for Load Rejection and Reduction

Discussing the start-up we have seen that an approx. 80% MCR-Bypass system is required for a supercritical unit. Experience and simulations have shown that on oil- or gas fired units a 80% MCR bypass with quick opening might be sufficient to avoid the MFT, but hardly on coal fired boilers with the lower rate of fire reduction.

The step in costs from a 80% to a 100% bypass system is relatively small compared to the advantages that are gained.

Fig. 7 shows the relative costs for the bypass valves vs. the size. The costs include HP-and LP-Bypass valves and 100% MCR RH safety valves for a supercriical 800 MW unit. For the 100%-case the additional costs to upgrade the HP-bypass to a combined HP-bypass safety system are shown also. For the 80% and 100% cases 4-valve solutions each for the HP-and LP-Bypass have been selected. This to get relatively small valves with a small wall thickness, avoiding the bypass valves becomeing a restricting element concerning thermocycling.

The 100% MCR Bypass offers the following advantages:
- Heat input and output of SH and RH can be kept in balance, i.e. metal temperature deviations and gradients remain within allowable limits.
- On once-through boilers operated in sliding pressure mode, the 100% - HP-Bypass allows a trip from part load without endangering the SH. The HP bypass is big enough to take over the produced steam without pressure increase, apart form short time dynamic effects. Without a bypass the pressure in the SH would have to be built up to the set pressure of the safety valves. During this time the SH and RH would have no flow, i.e. no cooling, which would lead to excessive metal temperatures. To avoid this a immediate MFT would be necessary. In case of coal fired units the pulverises can become the critical item for restart, because of the already mentioned necessary procedure.

- They safely avoid blowing of spring loaded safety valves.
- The 100% Bypass System allows operation of the boiler in real sliding pressure and at loads where full SH outlet temperature can be maintained. This helps to avoid forced cool down of the turbine HP-cylinder.

  When sizing the LP bypass for house load operation it comes out that it practically has to be a nominal 100% MCR bypass, independent of the size of the HP-Bypass.

  This because the max. flow of the valve is proportional to the inlet pressure.

  E.g. to obtain 25% LP-bypass flow the RH pressure must be 25% of nominal RH pressure and this is about what turbine suppliers accept for prolonged low load operation in respect of rotation loss in the HP cylinder. See Fig. 8.
- It has to be mentioned that with a bypass -independent of its size- during trip and cut back cold RH steam is available for deaerator, feedwater-tank, preheaters, etc.

  This helps to minimise temperature disturbances on the boiler inlet side and helps to stabilise boiler operation.
- The 100% MCR HP Bypass offers the possibility to transform it -with relative low additional costs- into a Combined Bypass and Safety System. This is accepted by the authorities of many countries among them PR of China and South-Korea.

  If this solution is accepted a number of additional operational and commercial advantages come into the game:

- The RH has flow during safety operation, i.e. an MFT is not necessary to prevent the RH from excessive metal temperatures. Disturbances on the boiler inlet side are kept low because RH steam is available for deaerator, preheaters etc.
- Loss of boiler water is kept to a minimum. Steam is not discharged to the atmosphere, but to the RH and from there through the LP-bypass to the condenser. The bigger the LP-Bypass capacity the smaller is the loss through the RH safety valves.
- External noise caused by blowing HP safety valves is avoided.
- Spring loaded safety valve have a tendency to become unstable. Fluttering and hammering can lead to damage. The very short stroking times can cause pressure shocks and high mass forces to the piping. The controlled and softer acting Combined Bypass and Safety Valves avoid these risks.
- Spring loaded HP-safety valves, their piping and noise silencers can be eliminated. Cost reduction for pipework and engineering --not including the spring loaded safety valves and silencers-- were estimated to be in the order of 750'000 to 1 Mio. US$ for a supercritical 500 MW unit.
- Spring loaded safety valves have a tendency to leak, due to erosion and/or the relatively low seating force.

  A rough estimation shows the order of magnitude of losses of fuel and water that can accumulate. A assumed leak with a cross section of 1 cm^2 causes a leak flow of approx. 9 t/h. If it is assumed that this leakage appears 1'000 h before the next revision then 9000 t of treated water and approx. 1300 t of coal would be wasted. With a assumed world market price for coal of 60 $/t and costs of only 10 $/t of treated water a total loss of 168'000 $ would result. If this happens 3 times in the life of the plant, then the losses become comparable with the costs for the HP bypass valves.

With Combined Bypass and Safety Valves the risk of leakage is much smaller because of the high seating force. Even if seat leakage would occur there would be minor losses only, because the leakage flows to the RH where energy and water are recuperated. A way to improve leakage of spring loaded safety valves is the addition of a pilot operated supplementary loading. Where these systems are allowed Combined Bypass and Safety Valve are allowed too. If a bypass system is provided anyhow then the extension to a Combined Bypass and Safety Valve offers more advatages and is the cheaper solution.

The high accuracy and the small hysteresis of the pressure switches together with the automatic and soft transfer from safety to bypass operation with automatic pressure control avoid the big drop of SH pressure as it would occur with spring loaded safety valves. For spring loaded safety valves the difference between set pressure and reclosing pressure is usually in the range of 10% of the set pressure. Providing the same set value for the safety function of both ty pes of valves, with the spring loaded valve the superheater pressure would drop below normal operating pressure. With the Combined Bypass and Safety Valve the SH outlet pressure stabilises on HP bypass set value, which is approx. 4% above normal operating pressure.(See Fig. 9). That means the pressure disturbance and the subsequent temperature gradients are smaller than with the spring loaded safety valves.

In addition to the above mentioned advantages Combined Bypass and Safety Valves seem to offer a higher reliability than spring loaded safety valves. This was shown by the German TÜV Rheinland, in a study (Ref. 1) in the frame of the European Standardisation. Fig. 10 shows the probabilities of the failure,that means not opening, of one and two safety valves of different types. The low failure probabilities of the pilot systems, mainly for hydraulic systems, prove that the system reliability is mainly given by the valve itself but not by its piIot system.It has to be mentioned that these results are not theoretical values but base on statistics of mandatory periodical safety valve tests.The author makes the reservation that the values for the spring loaded safety valves are valid for small valves only and expresses his hope that bigger valves might be better, but relevant statistics are not available.

3. SULZER-BYPASS CONTROLLER

30 years experience with Bypass-Systems, including controls and safety functions, show that the bypass-controller is neither an integrated part of the boiler controls nor the turbine controls. The bypass-system is rather an independent system which serves the operational requirements of the complete plant (i.e. boiler and turbine). The bypass controller ensures the life conserving operation of the heavily stressed valves and pipework, especially by means of accurate steam desuperheating in all operating modes and transients. It must allow the easy implementation of safety function for protection of the boiler, turbine and condenser.

3.1 HP-bypass-Controller

The Sulzer bypass-controller is an integrated system with signal conditioning, control and valve positioning functions. The type of operator interface can be tailored to the needs of the individual plant (i.e. desk insert or connection to a process video system). With a few standardised interface signals the Sulzer bypass-controller can be tied easily into an overall plant automation. The control function for start-up as well as for shut-down are fully automated.

Boiler start-up
The controller has to control and increase the SH-pressure according to the steam production of the boiler. The spray water injection controller has to control the temperature of the steam to the RH whenever steam is flowing through the bypass.

Turbine start-up
The HP-bypass-controller has to control the SH-pressure until the bypass is closed and the boiler master controller can take over the pressure control.

Load operation
The bypass is closed but the controller is ready to prevent excessive live steam pressure or excessive pressure gradients.

Turbine load rejection/trip
The controller opens the bypass valves, if necessary with the help of the quick opening devices, in order to prevent excessive live stream pressure and controls the pressure until the turbine picks up load again.

Safety Function:
In case the HP-Bypass is equipped with safety function this is functionally integrated into the bypass controller to ensure smooth transients between safety and control function.

3.1.1 Pressure Control

Figure 11 shows in detail the structure of the pressure controller and the pressure setpoint generator. The operating modes of the pressure controller are represented in the start-up diagram for a cold start of Figure 12.

At the begin of a cold start the minimum opening (Y_{min}) is active. It ensures immediately after ignition an open path and therefore a steam flow through the SH and RH.

When the valve position reaches a preadjusted value Y_m (determined by the desired steam flow during boiler start up) the setpoint generator begins to increase the pressure setpoint in accordance with the steam production of the boiler, but with a limited maximum gradient.

Once the target pressure for starting the turbine (P_{synch}) is reached, the setpoint generator switches to (fixed) pressure control. As the turbine starts to accept steam, the bypass will start to close until the turbine consumes all the steam produced by the boiler upon which the bypass is fully closed.

As soon as the bypass is closed the pressure setpoint tracks the actual pressure plus a threshold dp which keeps the bypass closed (follow mode). The maximum gradient of the pressure setpoint is still limited. If the life steam pressure exceeds this gradient, the bypass will start to open and the controller returns to pressure control mode. The pressure is controlled by the bypass until normal operation has been restored and the bypass is closed again.

When there is sufficient steam production to reach a predetermined minimum pressure (Pmin) the controller begins to control the live steam pressure by opening the bypass valves.

3.1.2 Temperature Control

Control of the steam temperature under all operating conditions requires a controller well matched to the wide range of operating conditions of a HP-bypass (low load, quick opening at full load, etc.). Accurate control of the temperature under all this operating conditions is an important life conserving factor for the heavily stressed walls of the valves and piping.

3.2 LP-bypass Controller

The LP-bypass controller operates independent of the HP-bypass controller. The dynamic behaviour of the LP-bypass control loop must be well adapted to the leading dynamic properties of the HP-bypass, mainly if there are safety or quick opening functions.

3.2.1 Pressure Control

The duty of the LP-bypass pressure controller for the different operating modes can be summarised as follows: (Fig. 13)

Boiler start-up
The controller has to control the steam pressure in the RH system. The injection controller has, whenever the LP-bypass is open, to control the desuperheating of the steam so that it can be accepted by the condenser.

Load operation
The bypass is closed but the controller monitors the rehear steam pressure in order to open the valves and control the pressure whenever an unacceptable pressure increase is monitored.

Whenever the condenser is not able to accept the steam or the injection water system is unavailable the controller has to close the bypass through a separate safe channel in order to protect the condenser.

During load operation the first-stage pressure of the turbine serves as load signal for the setpoint generator which generates a load dependent (sliding) pressure setpoint.

3.2.2 Injection Water Control

Because the steam conditions after the LP-bypass desuperheater are usually near or at saturation condition, the temperature after the desuperheater cannot be used as control signal. The necessary injection water flow and corresponding valve position of the injection valve must therefore be calculated from the steam flow and the steam conditions.

The steam flow is in turn a function of the steam conditions and the valve position of the bypass valve. The LP-bypass controller provides the necessary computing functions to calculate the necessary injection valve position from steam conditions and LP-Bypass valve position.

4. SOME VALVE DESIGN ASPECTS

Figures 14 and 15 show typical designs of HP and LP bypass valves. The valve body itself must not become a element that limits the temperature gradients of the boiler/turbine system. Therefore it must have a cyclic life as high as possible and be resistant against temperature shocks. The design principles to attain this are
-the use of forged materials
-the application of spherical shapes, which give the lowest wall thickness
-refined detail design, which avoids material accumulation, sharp edges and changes in cross sectional areas, and
-good surface finish.

This principles have been followed by our designer for years already and for advanced plants they become even of the higher importance. The experienc in the 600 °C unit Drakelow C learned us that also long-time creep deformation has to be taken in account.

An other Drakelow experience is the following: Valve stems that are exposed long time to 600 °C, even if they are made of complex alloyed 12% Cr-steel, build up a magnetite layer that leads to increase of diameter. In guides with small clearance this can cause seizing and subsequent malfunction of the valve. For the time being we are testing surface treatments that avoid this effect and don't increase manufacturing costs as strongly as e.g. stellit cladding would do.

One more point of attention are sealings and stuffing boxes. Graphite packings at temperatures above approx. 550°C suffer from heavy shrinkage due to oxidation with oxygen contained in the steam or in the air. Counter measures are metal shields on the graphite surfaces to avoid direct exposure to the oxygen or design measures to keep temperature below 550°C.

Fig. 14 shows a typical HP Bypass valve with integrated spray, designed for 212 bar and 580 °C. The „flow to open" design was selected to profit from the high opening forces for safety or quick opening functions. In addition this design principle offers following advantages:
-The high pressure part is of simple cylindrical shape without stress risers
-The more complex shapes with stress risers are on the low pressure side where smaller wall thickness is required.
-The stem with its guides and stuffing box and the cover seals are on the low pressure side where also the temperatures are lower, and therefore most of the above mentioned problems are not to be expected.

5. CONTRIBUTIONS OF THE BYPASS SUPPLIER TO PLANT AVAILABILITY AND FLEXIBILITY

To have optimum plant flexibility the maximum possible plant availability and consequently a high component availability is required. In other words on the one hand operating problems on components or subsystems which influence the availability must be eliminated as quickly as possible. On the other hand, if some damage is found the plant owner must be in a position to decide whether the component has to be repaired or replaced immediately or whether it can be used further and for how long, so that possibly a repair or exchange can be done during a regular and planned outage.

By means of two examples it is shown that the bypass supplier needs high technical competence and system knowledge to contribute effectively to this aspect of plant flexibility.

The first example refers to the Vestkraft Power Station on the Esbjerg harbour area in Denmark.

This plant is equipped with a 100% MCR HP Bypass System with safety function. It consists of 2 HP bypass valves, each with a capacity of 150 kg/s at 251 bar, 560°C, plus a set of spray valves.

As shown on Fig. 16, the design of the bypass piping system differs from usual designs. Per bypass valve, 2 pipes extend from the main steam pipe to a header. A sphere is arranged at one end of the header, and our bypass valve is seated directly on it. From the standpoint of stressing, a sphere with optimum shape seemed to be the right solution.

From the beginning of commissioning vibration and noise occurred in the bypass system. Measurements were taken immediately by I/S Elsamproject. Strong tonal vibrations and noise radiation in the range of 4 kHz were measured. Initial damages occurred on the measurement lines. Various correction measures were taken, but nothing helped.

Subsequently, the search for the cause was concentrated on the bypass valve, mainly initiated by an independent consultant of the customer. Despite the good experiences with this valve type on other installations we carried out a sequence of valve modifications beginning with slight modifications of seat and trim to drastic change of the principle of expansion and temporary elimination of the outlet cage to find out whether the change of the internal flow pattern would help.

No modification was successful, nor did any measurement allow the cause of the disturbance to be located. However, we became increasingly convinced that our first idea, that it might be a system problem, was correct. In the laboratories of Sulzer-Innotec and together with their experts we modelled the piping system upstream of the valve in a scale of 1:3 and simulated the behaviour of the installation.

The 4 kHz disturbance could not be reproduced but elimination of the sphere upstream of the bypass valve showed an impressive noise reduction.

Based on this we proposed modifying the inlet system as shown on Fig. 17. The customer agreed and modified the inlet header during an annual summer outage.

Restart on 23th August 1996 showed that the 4 kHz-Peak had disappeared, i.e. noise and vibration with the original valve are at an acceptable low level through practically the full valve stroke. (See Fig. 18).

The second example relates to HP-bypass valves with integrated spray system. A relative small number of these valves show cracking of the outlet part of the body after years. Usually the damage appears as a mesh of cracks of small depth, sometimes combined with few cracks of greater depth.

We feel the advantages offered by the integrated spray as
- compact design, low costs
- wide control range
- stable temperature and temperature distribution for the subsequent piping are so important that we want to maintain this design principle also for units with elevated temperature.

To give the plant owner minimal risk and maximum flexibility we have initiated three measures:
- A research program, combining experimental and numerical simulation to understand and improve spray systems
- A numerical tool that allows to decide with good reliability whether and how long a damaged valve can be operated further without repair. This gives a high flexibility in decision.
- A method for quick and effective repair.

The concept how the „residual life to repair" can be estimated is presented on Fig. 19. Real plant operating data together with valve design parameters allow the calculation of a „bounding crack propagation curve" (see Fig. 20).

The basis for this calculation is the penetration depth of temperature variations into the wall, whereby the amplitude and frequency are given by statistical distributions of turbulent flow. Elapsed operating time leads now to a theoretical crack depth. Measured crack depth allows calibration of the curve. The calculations show a relatively good match with the few practical examples available.

The allowable crack depth, found by fracture mechanics calculations, (R6-Method) together with the calibrated crack propagation curve defines now the further possible safe operating time between actual and allowable crack depth.

In addition possible requirements to be expected from authorities can be satisfied.

The residual life time to repair or replacement, the actual state of the plant and the required time for repair or replacement now allow a quick, safe decision how to proceed further.

We feel this method can lead to more freedom of decision in an uncomfortable situation and contribute to the flexibility of the plant.

Literature:

[1] Source: W. Bung, B. Föllmer:
Gesteuerte Sicherheitsventile in Kraftwerken gemäss deutschem Regelwerk
Bauarten, Anforderungen, Betriebserfahrungen
Stand der Übernahme in das europäische Regelwerk

[2] R. Rohner:
Sulzer Bypass-Systems for fossil fired power stations
E Mech I, Power Conf. 1988

SULZER THERMTEC

No.	SH-Outlet Conditions to match 500°C Turbine Temperature	Required SH-Flow in % MCR	Bypass Capacity in %MCR at Full Load Pressure of 250 bar
1	40 bar	21.5	135
2	60 bar	23.5	98
3	80 bar	25.0	78
4	100 bar	26.5	66

SUPERCRITICAL BOILER 250 BAR/600°C
TYPICAL BYPASS SIZES FOR WARMSTART

TABLE 1

Bypass Systems
IMechE May 97

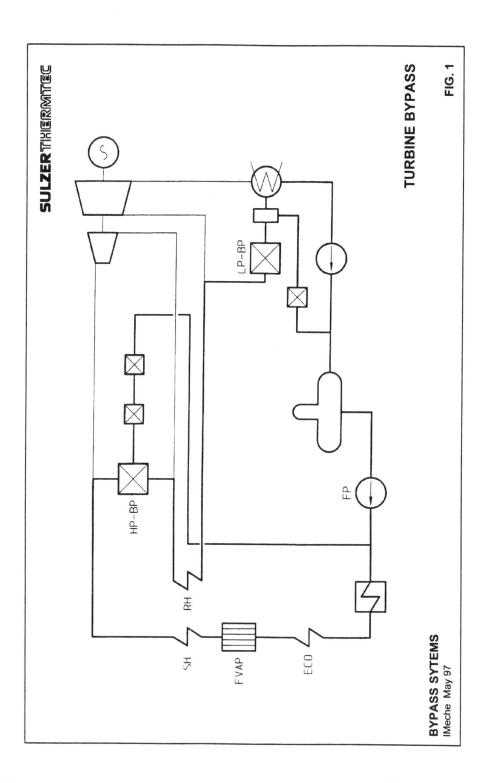

TURBINE BYPASS FIG. 1

BYPASS SYTEMS
IMeche May 97

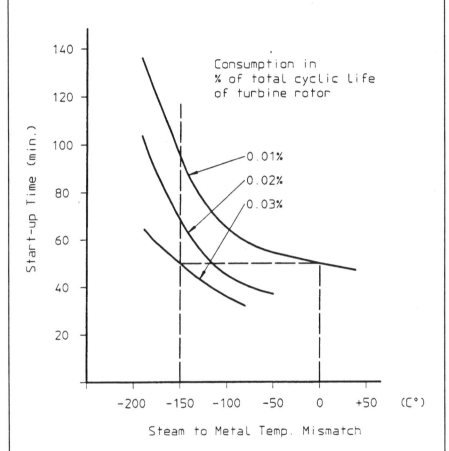

TURBINE START UP TIME VS: TEMPERATURE-MISMATCH

BYPASS SYSTEMS
IMechE May 97

FIG. 2

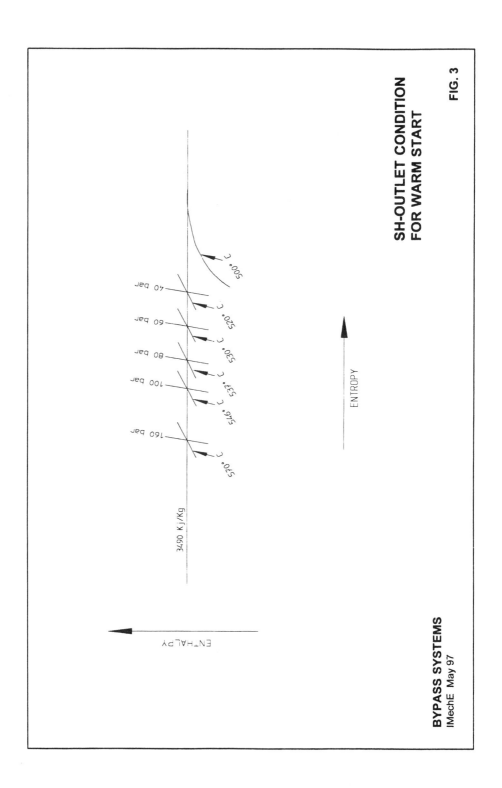

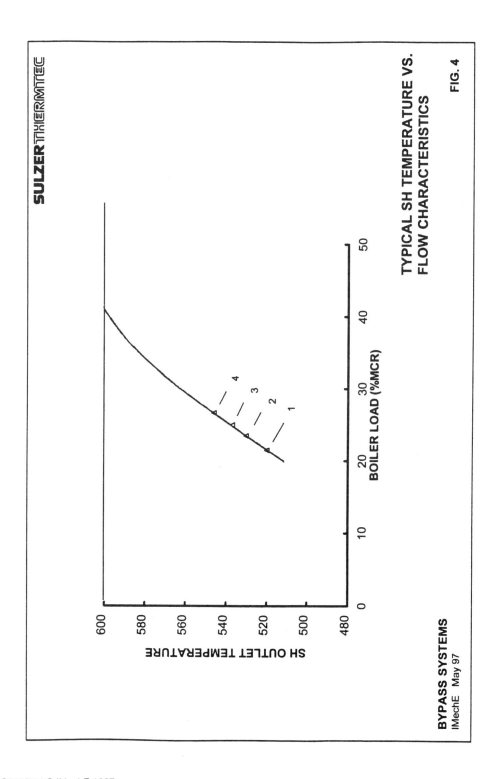

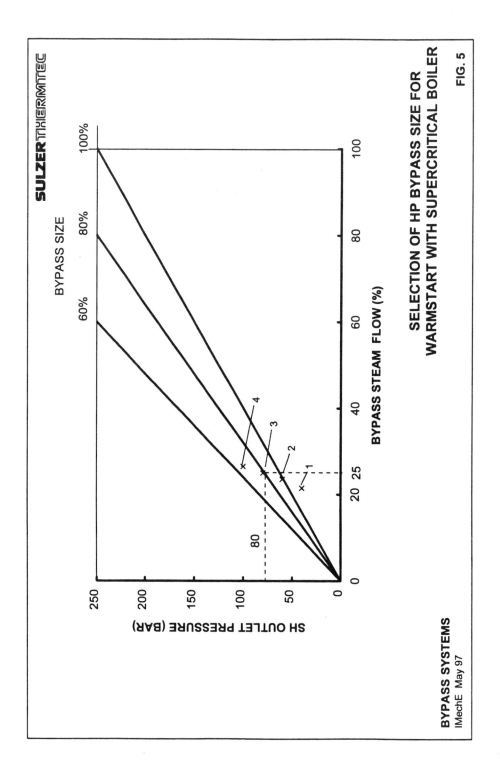

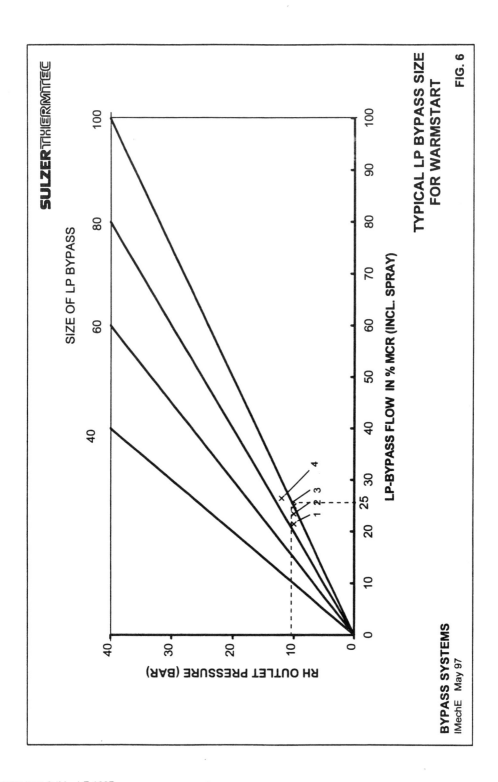

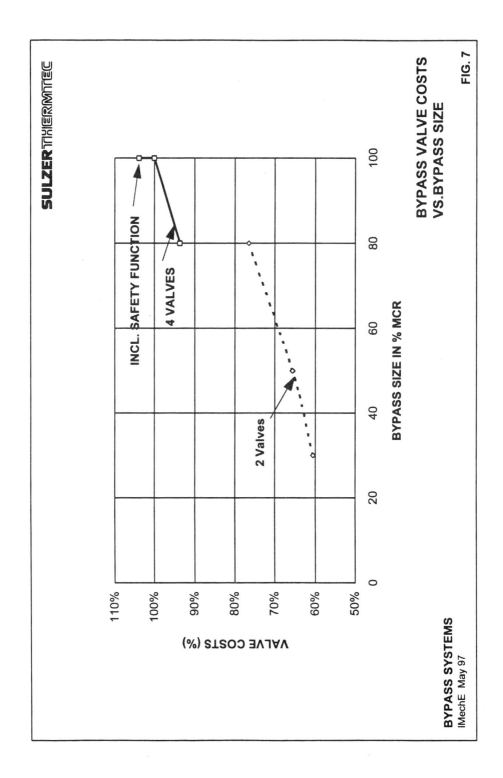

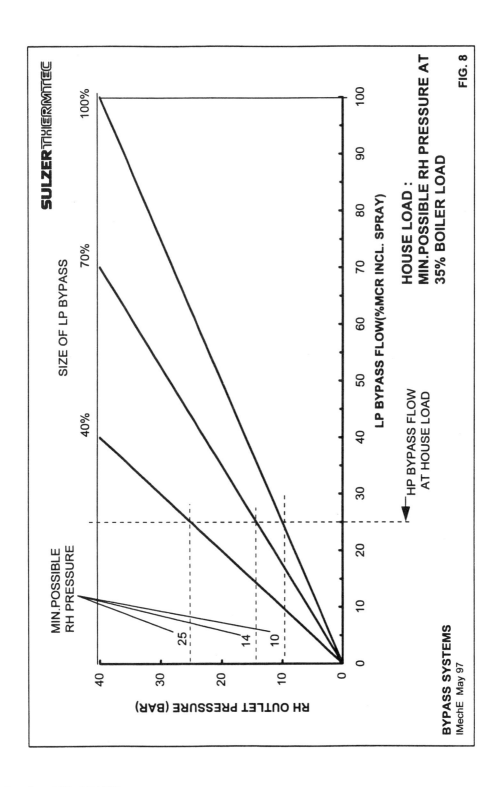

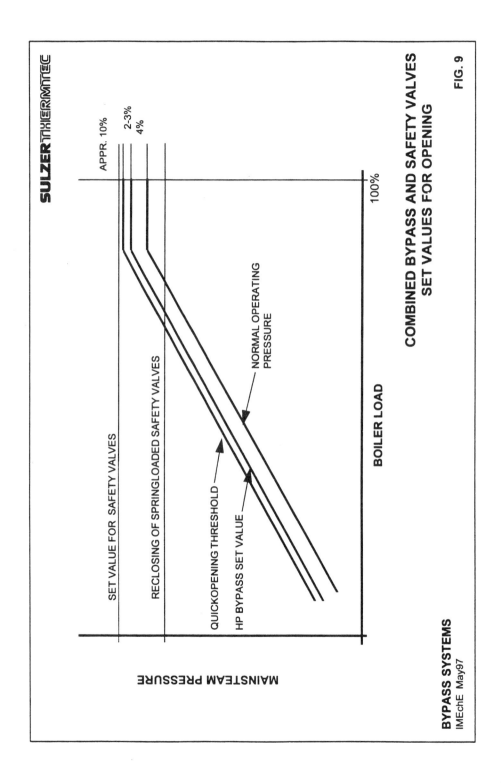

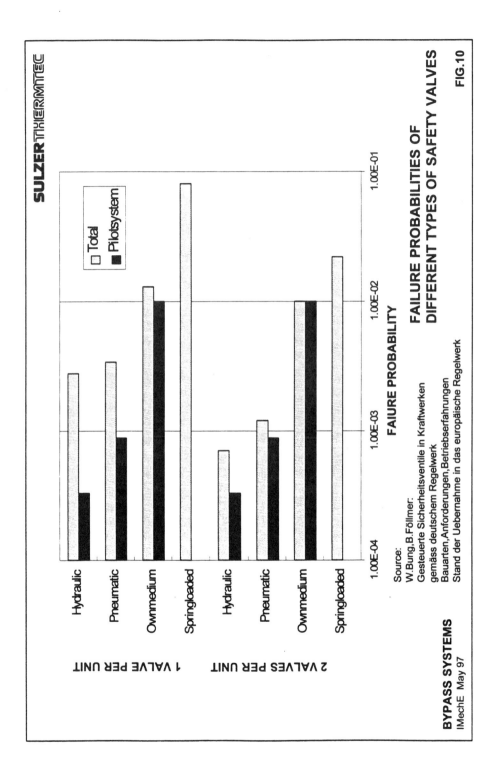

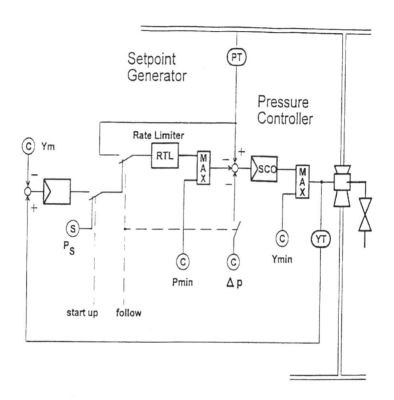

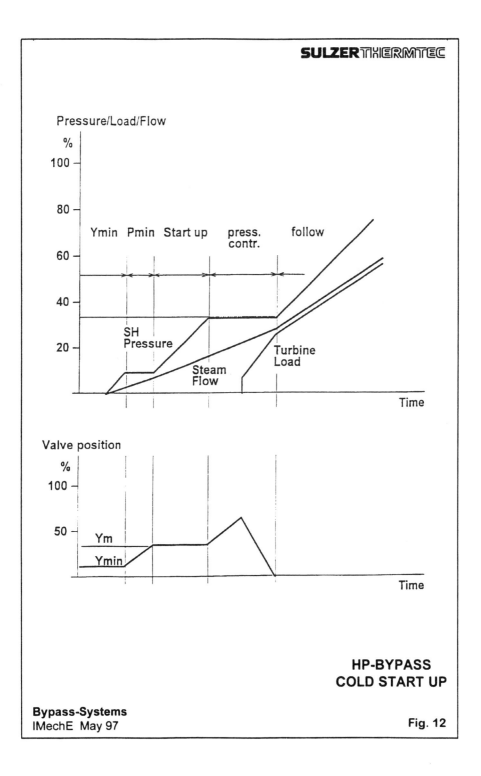

Fig. 12 HP-BYPASS COLD START UP

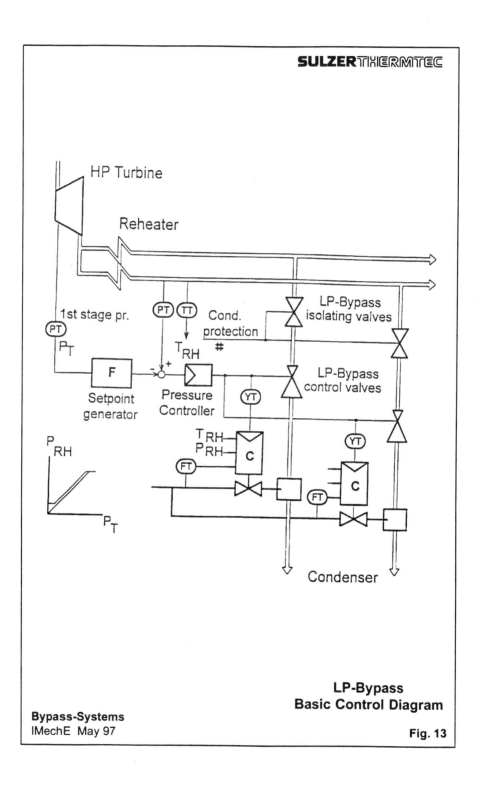

LP-Bypass Basic Control Diagram

Bypass-Systems
IMechE May 97

Fig. 13

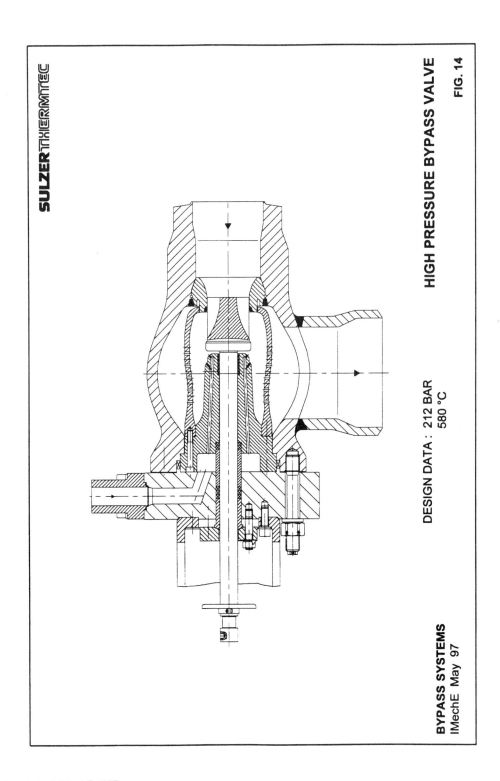

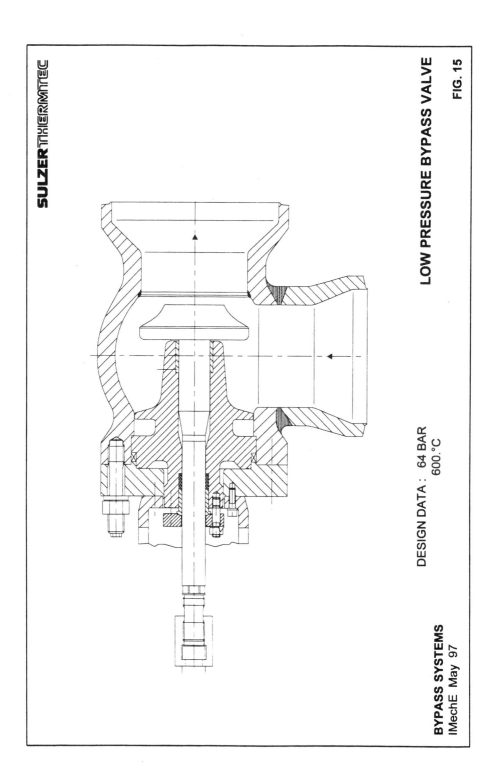

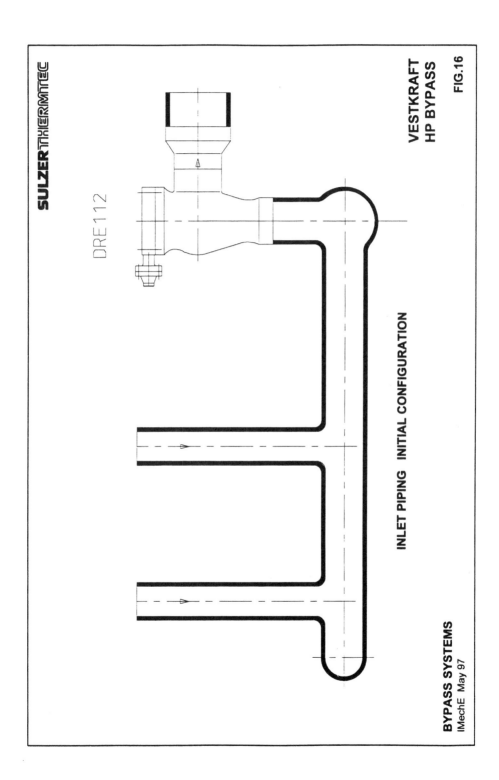

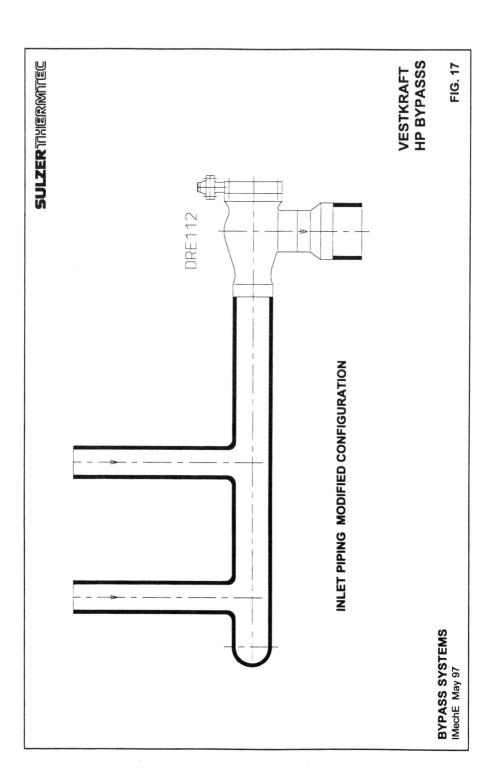

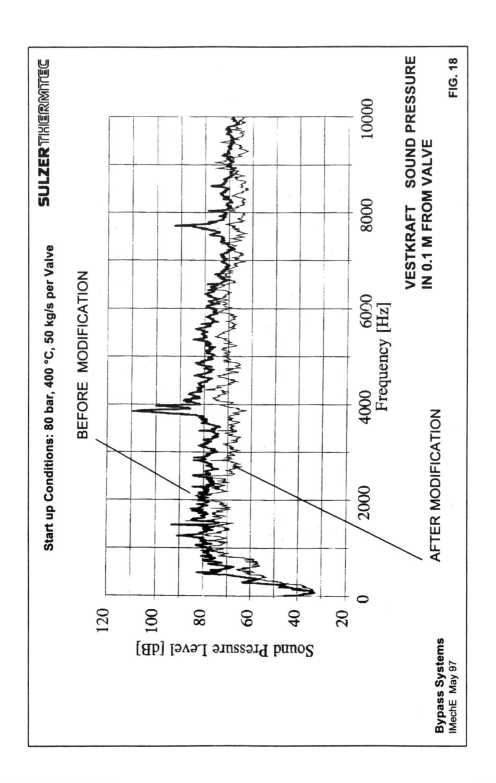

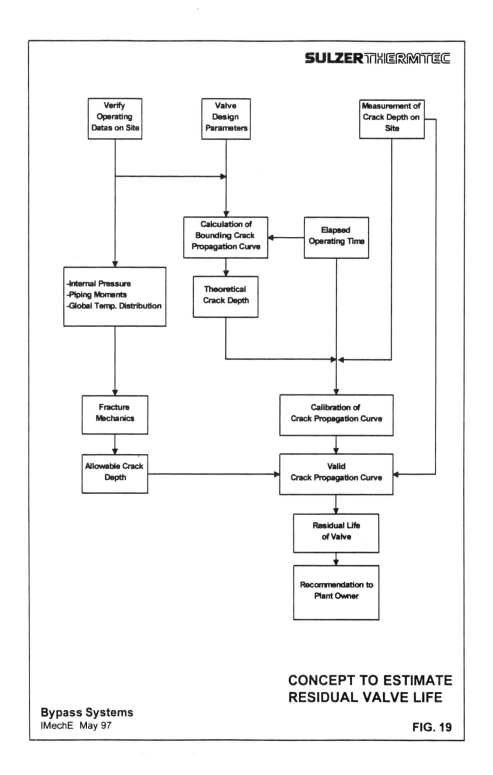

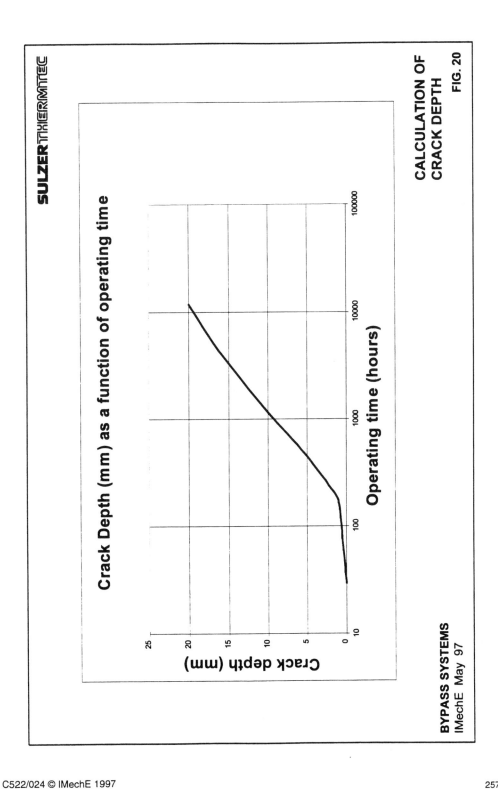

Authors' Index

Aregger, J 225–258

Bakker, W T 189–200
Bicego, V 201–210
Blum, H J R ....
Blum, R 3–17, 189–200

Blumer, U R 225–258
Bontempi, P 201–210
Bygate, R P 189–200

Deckers, M   65–84
Drosdziok, A 65–84

Franke, J 143–154
Fujita, T 87–98, 115–124
Fukui, Y 87–98

George, F 169–186
Gibbons, T B 189–200

Hald, J 3–16, 189–200

Hasegawa, Y 115–124
Hidaka, K 87–98

Kaneko, R 87–98
Kral, R 143-154

Lucon, E 201–210

Masuyama, F 189–200
Metcalfe, E 189–200
Morita, S 155–168

Naoi, H 115–124, 189–200

Oeynhausen, H 65–84
Ohgami, M 115–124

Paterson, A N 33–48

Price, S 189–200

Rohner, R R 225–258

Sakai, K 155–168
Sawaragi, Y 189–200
Scarlin, B 49–64
Scott, D H 213–224

Taylor, M 125–140
Taylor, N 201–210
Thornton, D V 125–140
Torkington, I R 169–186

Upton, M 169–186

Vanstone, R W 87–98

Wittchow, E 143–154